Guidebook to
Protein Toxins and Their Use
in Cell Biology

Guidebook to
Protein Toxins and Their Use in Cell Biology

Edited by

Rino Rappuoli

IRIS, Chiron Vaccines Immunobiological Research Institute in Siena, Siena, Italy

and

Cesare Montecucco

Centro CNR Biomembrane and Università di Padova, Padova, Italy

A SAMBROOK & TOOZE PUBLICATION
AT OXFORD UNIVERSITY PRESS
1997

#35758074

Oxford University Press, Great Clarendon Street, Oxford OX2 6DP

Oxford   New York

Athens   Auckland   Bangkok   Bombay   Buenos Aires
Calcutta   Cape Town   Dar es Salaam   Delhi   Florence   Hong Kong
Istanbul   Karachi   Kuala Lumpur   Madras   Madrid   Melbourne
Mexico City   Nairobi   Paris   Singapore   Taipei   Tokyo   Toronto

and associated companies in
Berlin   Ibadan

Oxford is a trade mark of Oxford University Press

Published in the United States
by Oxford University Press Inc., New York

A catalogue record for this book is available from the British Library

Library of Congress Cataloging in Publication Data
Guidebook to protein toxins and their use in cell biology/edited by
Rino Rappuoli and Cesare Montecucco.
(Sambrook and Tooze Guidebooks)
Includes bibliographical references and index.
1. Toxins–Handbooks, manuals, etc.   I. Rappuoli, Rino.
II. Montecucco, C. (Cesare)   III. Series: Guidebook series (Oxford, England)
QP631.G85   1997   615.9–dc21   96-29554

ISBN   0   19   859955   2   (Hbk)
ISBN   0   19   859954   4   (Pbk)

Typeset by EXPO Holdings, Malaysia

Printed in Great Britain by
The Bath Press

A computer system will be available from April 1997 to accompany *Guidebook to protein toxins and their use in cell biology*

Due to the rapid pace of biological research, the editors and publishers of this book believe it is important that its readers are kept informed of recent developments on these proteins. For this purpose, we have established a computer database that can be accessed through the worldwide web. This database will not include the full entries shown in this book; instead the authors have been asked to add, periodically, any new information on their protein that has been published since they wrote their original entry. Authors will be asked to deposit their updates from April 1997.

The update system can be accessed using any of the standard tools for browsing the worldwide web, such as Netscape or Mosaic. The URL for information relating to this book is *http://www.oup.co.uk/guidebooks/toxins*. For information on other Sambrook & Tooze Guidebooks, start from the Oxford University Press home page at *http://www.oup.co.uk/* and follow the links to the Guidebooks series.

Contents

Part 1. Membrane permeabilizing toxins

Part 2. Toxins affecting signal transduction

Part 3. Toxins affecting protein synthesis

Part 4. Cytoskeleton-affecting toxins

Part 5. Toxins affecting the immune and inflammatory response

Part 6. Toxins affecting membrane traffic

Part 7. Sodium channel targeted toxins

Part 8. Potassium channel-blocking toxins

Part 9. Calcium channel targeted toxins

Part 10. Acetylcholine receptor targeted toxins

Part 11. Ryanodine receptor Ca²⁺ channel toxins

Part 12. Presynaptic toxins

Part 13. Glutamate receptor targeted toxins

Contributors

Salah Abu-Raya, Department of Pharmacology, School of Pharmacy, Faculty of Medicine, The Adolph Weinberger Building, P.O.B. 12065 Jerusalem, Israel.

Michael E. Adams, Departments of Entomology & Neuroscience, University of California, 5419 Boyce Hall, Riverside, CA 92521, USA.

Abdu Adem, Department of Biochemistry, Biomedical Centre, Box 576, 751 23 Uppsala, Sweden.

Klaus Aktories, Institut für Pharmakologie und Toxikologie, Albert-Ludwigs-Universität Freiburg, Pharmakologisches Institut, Hermann-Herderstrasse 5, D-79194 Freiburg, Germany.

Joseph Alouf, Institut Pasteur de Lille, 1, rue du Professeur Calmette, B.P. 245, F 59019 Lille Cedex, France.

D. Barnett, School of Biological and Chemical Sciences, Deakin University, Geelong, Victoria 3217, Australia.

Baltazar Becerril, Dept of Molecular Recognition and Structural Biology, Instituto de Biotecnologia, Universidad Nacional Autonoma de Mexico, Apdo 510-3, Cuernavac, Morelos 62250, Mexico.

Jack A. Benson, Institute of Pharmacology, University of Zurich, Winterthurerstrasse 190, CH-8057 Zurich, Switzerland.

S. Bhakdi, Institute of Medical Microbiology, University of Mainz, Obere Zahlbacher Str. 67, D 55101 Mainz, Germany.

Allan L. Bieber, Department of Chemistry and Biochemistry, Arizona State University, Tempe, Arizona 85287-1604, USA.

Juan Blasi, Departament de Biologia Cellular i Anatomia Patològica, Universitat de Barcelona, Facultad de Medicina, Casanova 143, Barcelona 08036, Spain.

Eugenia Bloch-Shilderman, Department of Pharmacology, School of Pharmacy, Faculty of Medicine, The Adolph Weinberger Building, P.O.B. 12065 Jerusalem, Israel.

Cassian Bon, Unité des Venins, Institut Pasteur, 25 rue du Dr. Roux, 75724 Paris, France.

Patrice Boquet, Unité des Toxines Microbiennes, Institut Pasteur, 75724 Paris Cedex 15, France and INSERM U452 Faculté de Médicine de Nice, 06107 Nice Cedex 2, France.

Pierre E. Bougis, Laboratoire de Biochemie, Unité de Recherche Associée 1455, du Centre National de la Recherche Scientifique Institut Fédératif Jean Roche, Université de la Méditérranée, Faculté de Médicine Secteur Nord, 15 Bd. Pierre Dramard, F-13916 Marseille Cedex 20, France.

Chauncey W. Bowers, Division of Neurosciences, Beckman Research Institute of the City of Hope, 1450 East Duarte Road, Duarte, CA 91010, USA.

W. Dale Branton, Department of Physiology, Medical School, University of Minnesota, 6-255 Millard Hall, 435 Delaware Street S.E., Minneapolis, MN 55455, USA.

J. Thomas Buckley, Department of Biochemistry and Microbiology, University of Victoria, Box 3055, Victoria, BC, Canada V8W 3P6.

Lucio Cariello, Molecular Biology and Biochemistry Laboratory, Stazione Zoologica 'Anton Dohrn', Villa Comunale I, 80121 Naples, Italy.

Sabine Castano, Centre de Recherche Paul Pascal, CNRS, Avenue Schweitzer, 33600 Pessac, France.

Cristian Chiamulera, Department of Pharmacology, Glaxo Wellcome S.p.A. Medicines Research Centre, 37100 Verona, Italy.

Vincent A. Chiappinelli, Department of Pharmocology, George Washington University School of Medicine, 2300 Eye St. NW, Washington DC 20037, USA.

Francesco Clementi, Department of Medical Pharmacology, University of Milan, CNR Cellular and Molecular Pharmacology Center, Via Vanvitelli, 32, 20129 Milano, Italy.

R. John Collier, Department of Microbiology and Molecular Genetics, Harvard Medical School, 200 Longwood Avenue, Boston, Massachusetts 02115, USA.

A. Comis, School of Biological and Chemical Sciences, Deakin University, Geelong, Victoria 3217, Australia.

M. Connor, School of Biological and Chemical Sciences, Deakin University, Geelong, Victoria 3217, Australia.

Pascale Cossart, Unité des Interactions Bactéries-Cellules, Institut Pasteur, 28 rue du Docteur Roux, Paris 75015, France.

Lourdes J. Cruz, Marine Science Institute, University of The Philippines, Diliman, Quezon City 1101, The Philippines.

Gianfranco Donelli, Department of Ultrastructures, Istituto Superiore di Sanità, Viale Regina Elena 299, 00161 Rome, Italy.

Florian Dreyer, Rudolf-Buchheim-Institut für Pharmakologie der Justus-Liebig-Universität Giessen, Frankfurter Strasse 107, D-35392 Giessen, Germany.

Jean Dufourcq, Centre de Recherche Paul Pascal, CNRS, Avenue Schweitzer, 33600 Pessac, France.

Coumaran Egile, Unité de Pathogenie Microbienne Moléculaire, Institut Pasteur, 28 rue du Docteur Roux, 75724 Paris Cedex 15, France.

Christoph von Eichel-Streiber, Verfügungsgebäude für Forschung und Entwicklung, Institut für Medinzinische Mikrobiologie und Hygiene, 55111 Mainz, Germany.

Alessio Fasano, Gastrointestinal Pathophysiology Unit, Center for Vaccine Department, University of Maryland, School of Medicine, 10 South Pine Street, Baltimore, MD 21201-1192, USA.

Grazyna Faure, Unité des Venins, Institut Pasteur, 25 rue du Docteur Roux, 75724 Paris, France.

Francesco Ferraguti, Department of Pharmacology, Glaxo Wellcome Sp.A. Medicines Research Centre, 37100 Verona, Italy.

Mercedes Ferreras, CNR-ITC Centro Fisica Stati Aggregati, Via Sommarive 14, 38050 Povo, Trento, Italy.

Carla Fiorentini, Department of Ultrastructures, Istituto Superiore di Sanità, Viale Regina Elena 299, 00161 Rome, Italy.

Bernhard Fleischer, Bernhard Nocht Institute for Tropical Medicine, Department of Medical Micro- biology and Immunology, Bernhard-Nochtstrasse 74, D-20359 Hamburg, Germany.

Maria Rita Fontana, IRIS, Chiron Vaccines Immuno- biological Research Institute in Siena, Via Fiorentina 1, 53100 Siena, Italy.

Maria L. Garcia, Membrane Biochemistry & Biophysics Dept, Merck Research Laboratories, P.O. Box 2000, Rahway, NJ 07065-0900, USA.

S. Gasparini, CEA. Département d'Ingéniérie et d'Etude des Protéines, C.E Saclay, 91191 Gif/Yvette Cedex, France.

Maria Gasset, Instituto de Quimica Fisica Rocasolano, CSIC, Serrano 119, 28006 Madrid, Spain.

Werner Goebel, Lehrstuhl für Mikrobiologie, Biozentrum der Universität Würzburg, Am Hubland, D 97074 Würzburg, Germany.

Cecilia Gotti, CNR Cellular and Molecular Pharmacology Center, Via Vanvitelli, 32, 20129 Milano, Italy.

Eugene V. Grishin, Laboratory of Neuroreceptors and Neuroregulators, Shemyakin and Ovchinnikov Institute of Bioorganic Chemistry of Russian Academy of Sciences, Ul. Miklukho-Maklaya, 16/10, 117871 GSP- 7, Moscow V-437, Russia.

Franc Gubenšek, Department of Biochemistry and Molecular Biology, Jožef Stefan Institute, P.O. Box 3000, 1001 Ljubljana, Slovenia.

Ernst Habermann, Clinical Pharmacology, University Hospital, D-35385 Giessen, Germany.

Philip C. Hanna, Department of Microbiology, Duke University Medical Center, Durham, NC 27710, USA.

John B. Harris, Muscular Dystrophy Labs, Regional Neurosciences Centre, Newcastle General Hospital, Newcastle-upon-Tyne, United Kingdom.

B. M. Harrison, School of Biological and Chemical Sciences, Deakin University, Geelong, Victoria 3217, Australia.

Alan L. Harvey, Dept. Physiology and Pharmacology, Strathclyde Institute for Drug Research, University of Strathclyde, 204 George Street, Glasgow G1 1XW, Scotland.

Judit Herreros, Departament de Biologia Cellular i Anatomia Patològica, Universitat de Barcelona, Facultad de Medicina, Casanova 143, Barcelona 08036, Spain.

Randall K. Holmes, Department of Microbiology, Campus Box B-175, University of Colorado Health Sciences Center, 4200 East Ninth Avenue, Denver, CO 80262, USA.

M. E. H. Howden, School of Biological and Chemical Sciences, Deakin University, Geelong, Victoria 3217, Australia.

M. Hugues, Institut de Pharmacologie Moléculaire et Cellulaire, CNRS, 660 Route des Lucioles, Sophia Antipolis, 06560 Valbonne, France.

Julita S. Imperial, Department of Biology, University of Utah, Salt Lake City, Utah 84112, USA.

Mikael Jolkkonen, Department of Biochemistry, Biomedical Centre, Box 576, 751 23 Uppsala, Sweden.

Evert Karlsson, Department of Biochemistry, Biomedical Centre, Box 576, 751 23 Uppsala, Sweden.

M. Kehoe, Institute of Medical Microbiology, University of Mainz, Obere Zahlbacher Str. 67, D 55101 Mainz, Germany.

Richard A. Keith, Department of Bioscience, Zeneca Pharmaceuticals, 1800 Concord Pike, Wilmington, DE 19850-5437, USA.

William R. Kem, Department of Pharmacology & Therapeutics, College of Medicine, University of Florida, Gainesville, FL 32610-0267, USA.

Mikhail V. Khvotchev, Laboratory of Neuroreceptors and Neuroregulators, Shemyakin and Ovchinnikov Institute of Bioorganic Chemistry of Russian Academy of Sciences, Ul. Miklukho-Maklaya, 16/10, 117871 GSP-7, Moscow V-437, Russia.

Christine Kocks, Institute for Genetics, University of Cologne, Zulpicher Strasse 47, 50674 Köln, Germany.

Igor Križaj, Department of Biochemistry and Molecular Biology, Jozef Stefan Institute, P.O. Box 3000, 1001 Ljubljana, Slovenia.

Richard A. Lampe, Department of Bioscience, Zeneca Pharmaceuticals, 1800 Concord Pike, Wilmington, DE 19850-5437, USA.

Philip Lazarovici, Department of Pharmacology, School of Pharmacy, Faculty of Medicine, The Adolph Weinberger Building, P.O.B. 12065 Jerusalem, Israel.

Michel Lazdunski, Institut de Pharmacologie Moléculaire et Cellulaire, CNRS, 660 Route des Lucioles, Sophia Antipolis, 06560 Valbonne, France.

Emmanuel Lemichez, Unité des Toxines Microbiennes Institut Pasteur, 75724 Paris Cedex 15, France and INSERM U452 Faculté de Médecine de Nice, 06107 Nice Cedex 2, France.

Stephen H. Leppla, Laboratory of Microbial Ecology, National Institute of Dental Research, NIH, Bethesda, MD 20892, USA.

D. R. Lloyd, School of Biological and Chemical Sciences, Deakin University, Geelong, Victoria 3217, Australia.

Albrecht Ludwig, Lehrstuhl für Mikrobiologie, Biozentrum der Universität Würzburg, Am Hubland, D-97074 Würzburg, Germany.

Peter Maček, Department of Biology, University of Ljubljana, Večna pot 111, 1000 Ljubljana, Slovenia.

Antonio Malgaroli, DIBIT, Scientific Institute San Raffaele, Via Olgettina 58, 20132 Milano, Italy.

Marie-France Martin-Eauclaire, Laboratoire de Biochemie, Unité de Recherche Associée 1455, du

Centre National de la Recherche Scientifique Institut Fédératif Jean Roche, Université de la Méditérranée, Faculté de Médecine Secteur Nord, 15 Bd. Pierre Dramard, F-13916 Marseille Cedex 20, France.

J. Michael McIntosh, Department of Psychiatry, University of Utah, Salt Lake City, Utah 84112, USA.

Dietrich Mebs, Zentrum der Rechtsmedizin, University of Frankfurt, Kennedyyallee 104, D-60596 Frankfurt, Germany.

Jacopo Meldolesi, DIBIT, Scientific Institute San Raffaele, Via Olgettina 58, 20132 Milano, Italy.

Angela R. Melton-Celsa, Department of Microbiology and Immunology, Uniformed Services University of the Health Sciences, 4301 Jones Bridge Road, Bethesda, MD 20814, USA.

Gianfranco Menestrina, CNR-ITC Centro Fisica State Aggregati, Via Sommarive 14, 38050 Povo, Trento, Italy.

André Ménez, CEA Département d'Ingénierie et d'Etude des Protéines, C.E. Saclay, 91191 Gif/Yvette Cedex, France.

Cesare Montecucco, Centro CNR Biomembrane and, Università di Padova, Via Trieste, 75, 35121 Padova, Italy.

John R. Murphy, Evans Department of Clinical Research and Department of Medicine, Boston University Medical Center Hospital, Boston, MA 02118–2393, USA.

G. M. Nicholson, School of Biological and Chemical Sciences, Deakin University, Geelong, Victoria 3217, Australia.

P. Nicholson, School of Biological and Chemical Sciences, Deakin University, Geelong, Victoria 3217, Australia.

Timothy M. Norris, Departments of Entomology & Neuroscience, University of California, 5419 Boyce Hall, Riverside, CA 92521, USA.

Raymond S. Norton, Biomolecular Research Institute, 343 Royal Parade, Parkville, Victoria 3052, Australia.

Alison D. O'Brien, Department of Microbiology and Immunology, Uniformed Services University of the Health Sciences, 4301 Jones Bridge Road, Bethesda, MD 20814, USA.

Baldomero M. Olivera, Department of Biology, University of Utah, Salt Lake City, Utah 84112, USA.

Sjur Olsnes, Department of Biochemistry, Institute for Cancer Research, The Norwegian Radium Hospital, Montebello, 0310 Oslo, Norway.

M. Palmer, Institute of Medical Microbiology, University of Mainz, Obere Zahlbacher Str. 67, D 55101 Mainz, Germany.

Emanuele Papini, Centro CNR Biomembrane and, Università di Padova, Via Trieste, 75, 35121 Padova, Italy.

Liliana Pardo, Dept of Molecular Recognition and Structural Biology, Instituto de Biotecnologia, Universidad Nacional Autonoma de Mexico, Apdo 510-3, Cuernavac, Morelos 62250, Mexico.

J. A. Pearson, School of Biological and Chemical Sciences, Deakin University, Geelong, Victoria 3217, Australia.

Rossella Pellizzari, Dipartimento di Scienze Biomediche, Università di Padova, Via Trieste, 75, 35121 Padova, Italy.

Michael W. Pennington, Bachem Bioscience Inc., 3700 Horizon Drive, King of Prussia, PA 19406, USA.

Daniela Pietrobon, Centro CNR Biomembrane and, Università di Padova, Via Trieste, 75, 35121 Padova, Italy

Mariagrazia Pizza, IRIS, Chiron Vaccines Immuno-biological Research Institute in Siena, Via Fiorentina 1, 53100 Siena, Italy.

Olaf Pongs, Zentrum für Molekulare Neurobiologie, Institut für Neurale Signalverarbeitung, Martinstrasse 52–Haus 42, D-20246 Hamburg, Germany.

Michael R. Popoff, Unité des Toxines Microbiennes Institut Pasteur, 75724 Paris Cedex 15, France and INSERM U453 Faculté de Médicine de Nice, 06107 Nice Cedex 2, France.

Lourival D. Possani, Dept of Molecular Recognition and Structural Biology, Instituto de Biotecnologia, Universidad Nacional Autonoma de Mexico, Apdo 510-3, Cuernavac, Morelos 62250, Mexico.

Rino Rappuoli, IRIS, Chiron-Vaccines Immunobiological Research Institute in Siena, Via Fiorentina 1, 53100 Siena, Italy.

Holger Repp, Rudolf-Buchheim-Institut für Pharmakologie der Justus-Liebig-Universität Giessen, Frankfurter Strasse 107, D-35392 Giessen, Germany.

K. V. Retson, School of Biological and Chemical Sciences, Deakin University, Geelong, Victoria 3217, Australia.

Hervé Rochat, Laboratoire de Biochimie, Université d'Aix Marseille II, Boulevard Pierre Dramand, 13916 Marseille Cedex 20, France.

G. Romey, Institut de Pharmacologie Moléculaire et Cellulaire, CNRS, 660 Route des Lucioles, Sophia Antipolis, 06560 Valbonne, France.

Ornella Rossetto, Centro CNR Biomembrane and, Università di Padova, Via Trieste, 75, 35121 Padova, Italy.

Kirsten Sandvig, Department of Biochemistry, Institute for Cancer Research, The Norwegian Radium Hospital, Montebello, N-0310 Oslo, Norway.

Philippe J. Sansonetti, Unité de Pathogenie Micro-bienne Moléculaire, Institut Pasteur, 28, rue du Dr. Roux, 75724 Paris Cedex 15, France.

Giampietro Schiavo, Imperial Cancer Research Foundation, 44 Lincoln's Inn Field, London WC2A 3PX, UK.

H. Schweitz, Institut de Pharmacologie Moléculaire et Cellulaire, CNRS, 660 Route des Lucioles, Sophia Antipolis, 06560 Valbonne, France.

Peter Sebo,* Unité de Biochimie des Régulations Cellulaires, Institut Pasteur, 28 rue du Dr. Roux, 75015 Paris Cedex 15, France.

*Present address: Cell and Molecular Microbiology Division Institute of Microbiology of the Czech Academy of Sciences, Videnska 1083, CZ-142, Praha 4.

Denis Servent, CEA Département d'Ingénierie et d'Etude des Protéines, C.E. Saclay, 91191 Gif/Yvette Cedex, France.

Shoji Shibata, John A. Burns School of Medicine, Department of Pharmacology, University of Hawaii, 1960 East West Road, Honolulu, Hawaii 96822, USA.

Robert Slaughter, Membrane Biochemistry & Biophysics Dept, Merck Research Laboratories RY80N-31C, PO Box 2000, Rahway, NJ 07065-0900, USA.

I. Spence, School of Biological and Chemical Sciences, Deakin University, Geelong, Victoria 3217, Australia.

Douglas J. Steel, Department of Biology, University of Utah, Salt Lake City, Utah 84112, USA.

Fiorenzo Stirpe, Dipartimento di Patologia Sperimentale, Università di Bologna, Via S. Giacomo 14, 40126 Bologna, Italy.

John L. Telford, IRIS, Chiron Vaccines Immunobiological Research Institute in Siena, Via Fiorentina 1, 53100 Siena, Italy.

Susan Treves, Istituto di Patologia Generale, Università degli Studi di Ferrara, Via L. Borsari 46, 44100 Ferrara, Italy.

M. I. Tyler, School of Biological and Chemical Sciences, Deakin University, Geelong, Victoria 3217, Australia.

Agnes Ullmann, Unité de Biochimie des Régulations Cellulaires, Institut Pasteur, 28 rue du Docteur Roux, 75015 Paris Cedex 15, France.

A. Valeva, Institute of Medical Microbiology, University of Mainz, Obere Zahlbacher Str. 67, D 55101 Mainz, Germany.

Philip Washbourne, Centro CNR Biomembrane and, Università di Padova, Via Trieste, 75, 35121 Padova, Italy.

U. Weller, Institute of Medical Microbiology, University of Mainz, Obere Zahlbacher Str. 67, D 55101 Mainz, Germany.

H. I. Wilson, School of Biological and Chemical Sciences, Deakin University, Geelong, Victoria 3217, Australia.

Doju Yoshikami, Department of Biology, University of Utah, Salt Lake City, Utah 84112, USA.

Eliahu Zlotkin, Department of Cell and Animal Biology, Institute of Life Sciences, The Hebrew University of Jerusalem, Jerusalem 91904, Israel.

Francesco Zorzato, Istituto di Patologia Generale, Università degli Studi di Ferrara, Via L. Borsari, 46, 44100 Ferrara, Italy.

Preface

Poisons are chemical scalpels for the dissection of physiological processes
(Claude Bernard (1866) *Leçons sur les propriétés des tissus vivants*, p. 177, Paris)

Many prokaryotes, animals, and plants produce toxins that exert their toxic, sometimes lethal, activity toward other living organisms. In several cases, the same organisms produce more than one toxin at the same time. Toxins can be considered a sort of biological warfare that somehow 'helps' the toxin-producing species in their struggle for life. Toxins are believed to increase the chance of survival and/or proliferation and/or spreading of a particular organism in its particular environment. While it is simple to understand the advantages conferred by certain toxins contained in the snake or spider venoms, this is not always the case for plants or microorganisms. However, this is to be attributed to our present lack of knowledge of the ecology and evolutionary history of the toxin-producing species, rather than to a lack of 'significance' or 'usefulness' of the toxin.

Hence, natural toxins (to distinguish them from those designed and produced by humans) are the product of a long-term co-evolution of species sharing the same ecological niche. In this process, toxins have been shaped and modelled to selectively hit a key molecular target of the host or of the prey. Somehow, in the course of evolution, toxins have *'learned which are the most essential or the most vital processes of living organisms'*. Moreover, *'they have learned how to affect selectively the key molecule(s) of a given processes'*. In the light of neo-Darwinism, we can fully substantiate and appreciate the conclusion derived by Claude Bernard from his classical physiological studies on the action of healing and other toxic substances: toxins can be used as very precious chemical scalpels which help the biologist to reach processes and targets not accessible to the anatomist's scalpel. In its essence, this conclusion is still valid and it is the basis of the present volume in the Guidebook series.

■ Animal toxins

One may notice that all animal toxins are directed to a cell surface protein involved in an essential cell function. Nearly all of them are ligands of ion channels which are inactivated upon toxin binding, and interfere with neuronal or muscular functions (Parts 7–13). This is related to the fact that toxin-producing animals use venom to attack and immobilize the prey or the potential predator. Blocking the synapse, the specialized anatomical structure at the basis of the function of integrated organized animals, leads to the immobilization/paralysis of the injected animal. Perhaps owing to the need for rapid

action, these animal toxins are designed to act at the cell surface. Figure 1 summarizes the most important ion channels attacked by protein toxins.

■ Bacterial toxins

In this respect, bacteria are very different. They do not have such needs of rapidity of action. Moreover, they co-evolve with their hosts much more rapidly than animal species. Hence, they have been able to develop a variety of sophisticated strategies of survival and of modifications of host physiology in order to promote their own multiplication and spread. The study of these strategies has given rise to the emergence of a new discipline, which has been termed recently 'Cellular Microbiology'. Some of these strategies involve the use of protein toxins that attack the host cells in a variety of different ways. Figure 2 summarizes pictorially the fundamental aspects of a cell that are affected by bacterial toxins. A large group of toxins alters the plasma membrane integrity in such a way as to alter the membrane permeability barrier (Part 1). They do so with different mechanisms, but the end result is, in any case, the death of the cell. Membrane pore formation is being exploited in cell biology to gain access to the cell interior. Nearly all remaining bacterial protein toxins act inside the cell. These toxins usually consist of two parts: one involved in cell binding and penetration (termed B), and one domain (termed A) displaying intracellular enzymic activity. None of them enter the cell from the plasma membrane. Rather, they make use of the endocytic pathways elaborated by cells to intake ligands and nutrients and to digest away from the surface un-needed ligands and/or receptors, as well as bound materials. They bind to cells via the B part and are endocytosed, sometimes only at specialized portions of the cell surface, such as the apical membrane or the presynaptic membrane terminals. They are then taken up inside the lumen of intracellular compartments. Consequently, different toxins travel different cellular highways or minor roads, depending on the particular receptor they are bound to and on the cell type. They turn toxic only when they become translocation competent, i.e. they become capable of translocating domain A across the membrane into the cytosol. This is the least understood of the four steps of the cell intoxication process.

On the basis of the present knowledge, we can distinguish two group of bacterial toxins: those having a

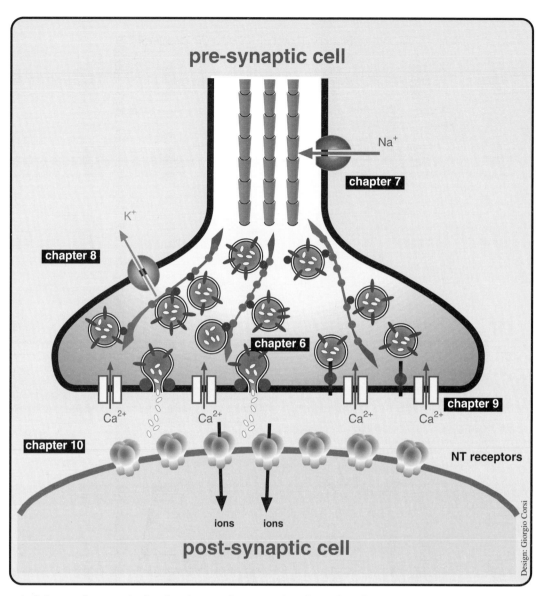

Figure 1. Scheme of a synapsis showing the most important ion channels and secretion patterns that are targets of protein toxins, and the chapters where they are described.

B part composed of two domains (diphtheria toxin, clostridial neurotoxins, etc.) and those having an oligomeric B domain (cholera toxin, shiga toxin, etc.). The first group of toxins enters the cell cytosol via an acidic intracellular compartment. There is evidence that low pH triggers a conformational change that makes these proteins able to insert in the lipid bilayer and to mediate the transfer of the catalytic A domain to the cytosol. The second group of toxins does not depend on low pH for their entry into the cytosol, and there is evidence that the

reduction of the single disulfide bond that links A to the B oligomer is sufficient to cause a change in the solubility properties of the A chain, so that it becomes able to partition into the lipid bilayer. Unfortunately, there is no information on the subcellular compartment where reduction takes place. Its identification would expand the use of these toxins as markers of intracellular reducing events. Once in the cytosol, these toxins interfere with a number of key cell events. Some toxins attack various components of the protein synthesis machinery (Part 3),

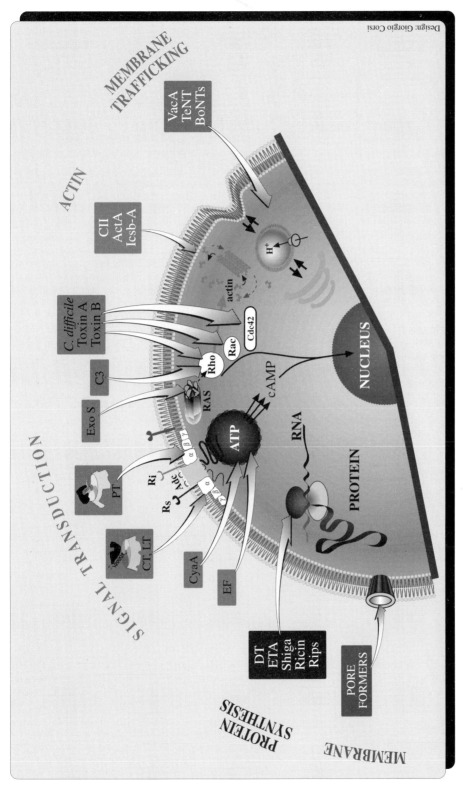

Figure 2. Schematic representation of bacterial toxins and their targets.

other toxins interfere in different ways with actin polymerization (Part 4). A group of toxins is specific for the trimeric G proteins that control several events of signal transduction or they increase directly the cellular c-AMP level (Part 2). Other toxins interfere with the immune and/or inflammatory response (Part 5), and another group of toxins affects different steps of vesicular trafficking inside cells (Part 6).

■ About this guidebook

Several toxins are well known in terms of their cellular activity, but their cellular targets remain to be identified (anthrax lethal factor, α-latratoxins, phospholipase snake toxins etc.). In general terms, the study of the mechanism of action of these toxins is important in two respects: (a) to understand the molecular pathogenesis of the disease in which the toxin is involved; (b) to learn about the molecular mechanism underlying the fundamental physiological process attacked by the toxin. Once this is known, the toxin can be used as a tool in the study of cell biology. This book focuses on the use of protein toxins as tools in cell biology. Accordingly, some toxins with poorly defined cellular effects are not included. At the same time, some toxins whose molecular targets are still to be identified, but which are expected to be very useful in the near future, are included. The aim of such a book is that of providing information and bibliographic references on the use of a large array of known protein toxins in the study of the molecular mechanisms involved

in the particular process considered. In some cases, the understanding of the mechanism of action of a toxin has provided novel therapeutic approaches to diseases affecting a particular function.

Natural toxins display an incredible variety of complexity, ranging from the simple formic acid of ants to bacterial proteins composed of thousands of amino acids. This book only considers protein or peptide toxins, essentially because of our restricted competence and to keep the volume within a reasonable size. We have organized the book in parts, which begin with an introduction to the cellular protein or process affected by a given group of toxins. Each part includes an entry for each different toxin, or group of toxins, with that particular target molecule (i.e. calcium channels) or cellular function (i.e. assembly of actin cables). The book does not mention all known protein toxins, rather, it considers them with respect to the interest of the cell biologists and groups toxins performing the same action within the same entry. Given the large number of known protein toxins and the fact that new ones are continuously being discovered, we may have failed to include appropriate entries. We apologize for such an inconvenience. We will be grateful to those who signal to us novel entries of potential interest to the cell biologist to be included in future editions of this Guidebook.

The Editors would like to acknowledge the editorial assistance of Catherine Mallia.

Siena and Padova,   R. R.
January 1997   C. M.

1

Membrane permeabilizing toxins

Introduction

Among the wide variety of toxins which bind and kill cells quite a large proportion have the plasma membrane as a target. Whatever their origin, structure, and size, varying from small peptides to large sophisticated oligomeric proteins, all perform a similar function: they permeabilize cells. The action of all the toxins assembled in this part of the book results from a physical interaction with membrane components. They are defined as direct lytic factors as opposed to other toxins having enzymatic properties and able to chemically degrade the membrane.

The first primordial function of the plasma membrane is to maintain all materials, from macromolecules to small solutes, selectively in the cell, i.e. the membrane is equipped in order to very selectively and efficiently regulate permeability. Consequently, it is not surprising that many bacterial toxins act on the plasma membrane and have evolved to change efficiently the first critical property required for living. Indeed, any unbalanced entry, or efflux of ions or small molecules should be opposed by selective pumps and channels in order to recover, as soon as possible, the physiological conditions by active transport. But if the nonselective fluxes induced by toxins overcome the capabilities of the cells to recover their equilibrium quickly, due either to an important transitory change, 'a burst', or to a permanent leakage through a 'hole', cells will generally die. This often results from the entry of ions inducing metabolic disorders, like Ca^{2+} entry, causing entry of water which swells the cell in order to compensate the osmotic pressure due to the presence of concentrated polyelectrolytes inside the cell. Conversely, the escape of vital compounds, again ions and small metabolites, can be lethal. Then all the following toxins have the same effect, they only differ in the way they perform it, in their efficiency in inducing the deleterious imbalance of permeability, and in the detailed structure and mechanism causing an open structure through which lethal leakage will occur.

In order to generate defects, channels, or holes in the membrane, protein toxins should provide a pathway for water and small solutes. Whatever the detailed structures, all such proteins must be amphipathic: one part should be lining a water filled domain, whereas others should be in contact with lipid chains and/or apolar segments of integral membrane proteins. This amphipathicity can be achieved at different levels:

1. By secondary amphipathic structural segments like helices or β-sheets, as for the peptidic toxins (see entry for δ-toxin p. 13) or porins (Weiss and Schulz 1992).Toxins are dominated by the requirement of being amphipathic.The peptidic toxins are more soap-like, weakly specific, self-associated both in buffer and in the membrane and can even solubilize lipids by forming reversed disc-like structures (Cornut *et al.* 1993; Saberwal and Nagaraj 1994).

2. By tertiary amphipathic structures for sophisticated Janus type proteins. The constituting amphipathic helices are packed with an apolar core when in buffer, and with an hydrophilic one when embedded into lipids (Van der Goot *et al.* 1991).

3. By quaternary amphipathic structures of proteins, which oligomerize in buffer by burying their apolar faces, and conversely oligomerize in the membrane with a polar hole or central channel. A variant of such a situation was recently proposed from the X-ray structure of annexins (Luecke *et al.* 1995). The aerolysin structure provides the first example of a β-sheeted oligomeric channel, whose oligomerization takes place only in the membrane (Parker *et al.* 1994).

Increasing the size of toxins generally allows an increase in the selectivity for a more unique target, either a lipid such as SLO and thiol-activated toxins (see corresponding entries), or a defined protein as documented for α-toxin (Walker *et al.* 1992). This also allows water solubility to be maintained by burying the membrane active area and avoiding too severe aggregation in buffer, which will decrease the affinity and availability of the toxin for its target.

Whatever the toxin, different steps which govern their cytotoxicity can be identified:

1. Toxins should interact selectively or be able to cross the thick glycosylated outer part of most cells.

2. Then toxins quickly bind to a selected membrane protein and/or lipids with high affinity. This results generally in an increased concentration of toxins more or less reversibly adsorbed on to the membrane. Most of the hydrophobic effects are already recovered and conformational changes take place concomitantly.

3. Toxins then reorient and penetrate into the membrane. Such a step is governed by the membrane's properties, it is therefore strongly dependent on the physical state of the lipids: the more defects in packing, the more easily toxins penetrate. This step is slow, it can take minutes to form a new structure (Yianni *et al.* 1986), and is strongly temperature-dependent as is well documented for SH-activated toxins (Alouf and Geoffroy 1991).

4. Embedded toxins oligomerize differently or aggregate in order to expel the polar or charged domains from lipid contact. This generally requires lateral diffusion, i.e. fluid membranes. For the larger toxins quite specific interactions occur between the subunits leading to a single well-defined structure and properties like those of α-toxin (Ostolaza et al. 1993). But oligomerization could be rather nonselective and lead to high molecular weight aggregates as in the case of SH-activated toxins (see entries for SLO p. 5 and others). The progressive change of the channel size is also documented for the peptidic toxins (see entry for δ-toxin, p. 13). Crosslinking of such toxins drastically increases their effects (Vogel et al. 1995).

Most of the toxins are produced and/or stored in a protoxin inactive form. The activation step varies, it could be a cleavage of an N_{term} acidic peptide like for melittin, or a C_{term} proteolysis as in alveolysin, but the *Escherichia coli* hemolysin activation results from an acylation. All such steps increase the affinity for the membrane which appears to be essential for activity. High cytotoxicity and high affinity for the membrane seem to be correlated, for instance the cytotoxic peptides have to be present in 10^5 to 10^6 copies in the target cell to get lysis, their LD_{50} are rather high 10^{-8} to 10^{-7} M, their nonstereo-specific binding is governed by the water/membrane partition coefficient (Fisher et al. 1994; King et al. 1994). In contrast, thiol-activated toxins have a higher affinity and clear stereospecificity for cholesterol (see entry p. 7). Indeed, one of the highest affinities is that of *E. coli* hemolysin, which has probably the highest selectivity and then efficiency (see entry p. 18).

At sublytic concentrations most of the toxins already induce changes in many cellular properties. Such effects can be defined as secondary, and are generally accounted for by an increased permeability with special effects due to Ca^{2+} entry. This results in production of eicosanoid compounds through activation of phospholipases, secretions, or inhibition of other enzymes. Moreover, the best documented synergistical activation of lipases implies direct interactions between toxins and the substrate and products of the reaction, but until now no direct toxin–enzyme contacts have been established. This leads to the synergism between the direct lytic factors described in this part and the indirect lytic ones which are for example inflammatory toxins (see Part 5) and signal transductory ones (see Part 2).

Finally, despite quite impressive progress being made on the structures and modes of action of the toxins, we do not as yet understand the molecular laws for the cell selectivity out of a few significant cases like cholesterol-sensitive toxins. Generally, we do not have any clear explanation for increased sensitivity of some eukaryotic cells and the way membrane from mammals, or bacteria, responds differently from species to species to the same toxin attack. The comparison of toxins and physiological channels will also highlight the differences and common features (Bladon et al. 1992).

References

Alouf, J. E. and Geoffroy, C. (1991). The family of antigenically related cholesterol-binding, sulphydryl-activated, cytolytic toxins. In *Bacterial protein toxins* (ed. J. E. Alouf and J. H. Freer), pp. 147–86, Academic Press, London.

Bladon, C. M. , Bladon, P., and Parkinson, J. A. (1992). δ-Toxin and analogue as peptide models for protein ion channels. *Biochem. Soc. Trans.*, **92**, 862–4.

Cornut, I., Thiaudiere, E., and Dufourcq, J. (1993). The amphipathic helix in cytotoxic peptides. In *The amphipathic helix* (ed., R. M. Epand), pp. 173–219, CRC Press, Boca Raton, Florida.

Fisher, T. J., Prendergast, F. G., Ehchardt, M. R., Urbauer, J. L., Wand J., Sedarous, S. S., et al. (1994). Calmodulin interacts with amphipathic peptides composed of all D-amino acids. *Nature*, **368**, 651–3.

King, T. P., Wade, D., Coscia, M. R., Mitchell, S., Kochoumian, L., and Merrifield, B. (1994) Structure–immunogenicity relationship of melittin, its transposed analogues and D-melittin. *J. Immun.*, **152**, 1124–1131.

Luecke, H., Chang, B. T., Mailliard, W. S., Schlaepfer, D. D., and Haigler, H. T. (1995). Crystal structure of the annexin XII hexamer and its implications for bilayer insertion. *Nature*, **378**, 512–15.

Ostolaza, H., Bartholome, B., Ortiz de Zarate, I., de la Cruz, F., and Goni, F. M. (1993). Release of lipid vesicle contents by bacterial protein toxin α-haemolysin. *Biochim. Biophys. Acta*, **1147**, 81–8.

Parker, M. W., Buckley, J. T., Postma, J. P. M., Tucker, A. D., Leonard, K., Pattus, F., et al. (1994). Structure of Aeromonas toxin proaerolysin in its water soluble and membrane-channel states. *Nature*, **367**, 292–5.

Saberwal, G. and Nagaraj, R. (1994). Cell-lytic and antibacterial peptides that act by perturbing the barrier function of membranes: facets of their conformational features, structure–function correlations and membrane-perturbing abilities. *Biochim. Biophys. Acta*, **1197**, 109–31.

Van der Goot, F. G., Gonzàles-Mañas, J. M., Lakey, J. H., and Pattus, F. (1991). A 'molten globule' membrane insertion intermediate of the pore-forming domain of colicin A. *Nature*, **354**, 408–10.

Vogel, H., Dhanapal, B., Pawlak, M., Meseth, U., Eggleston, I. M., and Mutter, M. (1995). Membrane channel forming properties of melittin-TASP. In *Peptides 1994*, (ed. H. S. L Maia) pp. 449–50, ESCOM, Leiden.

Walker, B., Krishnasastry, M., Zorn, L., Kasianowicz, J., and Bayley, H. (1992). Functional expression of the α-hemolysin of *Staphylococcus aureus* in intact *Escherichia coli* and cell lysates. *J. Biol. Chem.* **267**, 10902–9.

Weiss, M. S. and Schulz, G. E. (1992). Structure of porin refined at 1.8 Å resolution. *J. Mol. Biol.*, **227**, 493–509.

Yianni, Y. P., Fitton, J. E., and Morgan, C. (1986). Lytic effects of melittin and δ-lysin from *Staphylococcus aureus* on vesicles of DPPC. *Biochim. Biophys. Acta*, **856**, 91–8.

■ *Jean Dufourcq and Sabine Castano:*
Centre de Recherche Paul Pascal, CNRS,
Avenue Schweitzer,
33600 Pessac,
France

Streptolysin (*Streptococcus pyogenes*)

SLO is a 61 kDa protein produced by Streptococcus pyogenes *A and C. SLO binds specifically to cholesterol in target membranes. Bound toxin molecules associate with each other to form large, arc- and ring-shaped polymers that insert into the bilayer to produce transmembrane pores of up to 35 nm diameter.*

Streptolysin O (SLO) is produced by beta-hemolytic strep-tococci. It belongs to the group of thiol-activated toxins, a large family of homologous cytolysins which are secreted by gram-positive bacteria and share extensive homology (Alouf and Geoffroy 1991). All of these toxins bind to cell membranes containing cholesterol and then polymerize to form pores of up to 35 nm diameter (Bhakdi *et al.* 1985). The largest stretch of homology among all known thiol-activated toxins is situated about 40 amino acids upstream of the C-terminus and contains the single cysteine residue of the molecule. Chemical modification of this cysteine residue abrogates the ability of the toxins to bind to cell membranes (Iwamoto *et al.* 1987); thus the C-terminus plays an essential role in binding. At present, no other features of toxin function could be clearly assigned to parts of the protein sequence. The SLO gene has been cloned and sequenced (Kehoe *et al.* 1987; data bank accession numbers: M18638, P21131).

Streptococcal cysteine-protease cleaves the native, 61 kDa toxin close to its N-terminus to yield a 55 kDa

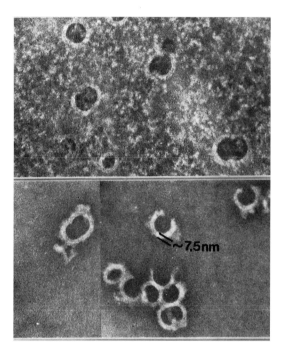

Figure 2. Top: Negatively stained erythrocyte membrane lysed by streptolysin O (SLO) showing curved rods (25–100 nm long and approximately 7.5 nm wide with inner radius of curvature of 16–18 nm. Most rods are approximately semicircular, often joined in pairs at their ends. Dense accumulations of stain are seen at the concave side of the rods. Bottom: Negative staining of isolated SLO oligomers, showing numerous curved rod structures identical to those found in toxin-treated membranes.

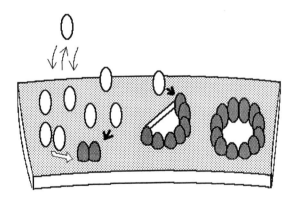

Figure 1. Schematic representation of streptolysin O oligomerization.

→: reversible association of monomeric streptolysin with membranes containing cholesterol.

⇒: nucleation step, two monomers react to form a stably membrane-embedded start complex.

→: oligomerization proceeds by successive addition of monomers, giving rise to arc- and finally ring-shaped complexes (consisting of 25 of 50 monomers).

truncate with unaltered pore-forming activity (Pinkney *et al.* 1995). The 61 kDa and 55 kDa toxin forms are present in streptococcal culture supernatants. Binding of SLO occurs very rapidly and in an essentially non-saturable fashion even at low temperature and without ionic requirements. After a rate-limiting nucleation step of second order, oligomerization proceeds by successive addition of monomers (Palmer *et al.* 1995). Electron microscopy reveals arc-shaped intermediate stages spanned by a free edge of membrane lipids (Bhakdi *et al.*

1985), suggesting that transmembrane pores arise concomitant with oligomer growth. The strong dependence of this process on temperature and toxin concentration allows for good control of experimental cell permeabilization.

■ Purification and sources

SLO can be isolated from streptococcal culture supernatants by ion exchange chromatography (Bhakdi et al. 1984) and chromatography on SH-Sepharose (Alouf and Geoffroy 1988). The purification product usually comprises native and N-truncated SLO, both forms display the same functional activity. A protocol for obtaining recombinant SLO from E. coli in fusion with maltose-binding protein has been established (Weller, manuscript in preparation). The commercial products may contain considerable amounts of impurities including proteases. Purified SLO is stable when stored lyophilized at −70 °C.

■ Toxicity

Functional activities are most conveniently assessed by determining the hemolytic titer against rabbit erythrocytes. Tests should be performed in the presence of 2 mM DTT and 0.1 per cent albumin. A toxin concentration of approximately 2–3 ng/ml causes 50 per cent lysis of a rabbit erythrocyte suspension (1.25×10^8 cells/ml). The sensitivity of nucleated cells towards SLO varies (Walev et al. 1995), as do the domains of polarized cells. For example, the apical side of MDCK cells is approximately 10 times more resistant than the basal side (Pimplikar et al. 1994). The cause of these varying sensitivities is largely unknown but may be due to formation of nonfunctional polymers (Walev et al. 1995). Functional activity in nucleated cells is most conveniently tested by determining the toxin concentration required to render the majority of cells trypan-blue or propidium iodide-positive. Measurements of LDH release or of ATP-depletion may also be employed.

SLO does not damage intact skin or mucous membranes. Use of the toxin hence does not present problems to operators. Systemic toxic reactions and organ dysfunction are induced upon intravasal administration or local application to organs. The LD_{50} for mice is 8–25 μg/kg (Smyth and Duncan 1978). All adults have antibodies against SLO that neutralize its activity in the application range of < 1 μg/ml (Aguzzi et al. 1988; Falconer et al. 1993).

■ Use in cell biology

SLO is being widely used to produce large pores in cell membranes, allowing the introduction of macromolecules to the cytoplasm (Ahnert-Hilger et al. 1989; Bhakdi et al. 1993). When applied to intact cells, SLO will be quantitatively absorbed to cholesterol and trapped in the plasma membrane; hence, intracellular membranes are spared (Esparis Ogando et al. 1994). The most reliable way to ensure a selective action on the plasma membrane is to expose cells to the toxin at low temperature; binding occurs rapidly and virtually quantitatively, whereas polymerization is retarded (Alouf and Geoffroy 1991; Rapp et al. 1993). Thereafter, permeabilization is effected by raising the temperature. At this stage, macromolecules that are to be applied to the cytoplasm (e.g. antibodies) can be added to the medium. When cells are suspended in appropriate media, a variety of machineries continue to operate over quite a prolonged time period. Examples include trafficking of membrane vesicles, and import of proteins and peptides into peroxisomes (Wendland and Subramani 1993) and into the endoplasmic reticulum (Tan et al. 1992; Momburg et al. 1994). In the future, it will probably become possible to permeabilize intracellular membranes, e.g. by microinjecting SLO into cells. A recent report, yet to be confirmed, contends that it is also possible to transiently permeabilize cells, allowing introduction e.g. of antisense oligonucleotides under retention of viability (Harvey et al. 1994). Finally, polymerization-defective SLO mutants are being constructed that may become useful as noncytotoxic probes for localizing membrane cholesterol.

■ References

Aguzzi, F., Rezzani, A., Nespoli, L., Monafo, V., and Burgio, G. R. (1988). Anti-streptolysin O titer, fifty-five years after Todd: a reappraisal of its clinical significance. Boll. Ist. Sieroter. Milan., 67, 162–4.

Ahnert-Hilger, G., Bader, M. F., Bhakdi, S., and Gratzl, M. (1989). Introduction of macromolecules into bovine adrenal medullary chromaffin cells and rat pheochromocytoma cells (PC12) by permeabilization with streptolysin O: inhibitory effect of tetanus toxin on catecholamine secretion. J. Neurochem., 52, 1751–8.

Alouf, J. E. and Geoffroy, C. (1988). Production, purification and assay of streptolysin O. In Microbial toxins: tools in enzymology (ed. S. Harshman), pp. 52–9, Academic Press, San Diego.

Alouf, J. E. and Geoffroy, C. (1991). The family of antigenically related, cholesterol-binding ('sulphhydryl-activated') cytolytic toxins. In Sourcebook of bacterial protein toxins. (ed. J. E. Alouf and J. H. Freer), pp. 147–86, Academic Press, London.

Bhakdi, S., Roth, M., Sziegoleit, A., and Tranum-Jensen, J. (1984). Isolation and identification of two hemolytic forms of streptolysin-O. Infect. Immun., 46, 394–400.

Bhakdi, S., Tranum-Jensen, J., and Sziegoleit, A. (1985). Mechanism of membrane damage by streptolysin-O. Infect. Immun., 47, 52–60.

Bhakdi, S., Weller, U., Walev, I., Martin, E., Jonas, D., and Palmer, M. (1993). A Guide to the use of pore-forming toxins for controlled permeabilization of cell membranes. Med. Microbiol. Immunol., 182, 167–75.

Esparis Ogando, A., Zurzolo, C., and Rodriguez Boulan, E. (1994). Permeabilization of MDCK cells with cholesterol binding agents: dependence on substratum and confluency. Am. J. Physiol., 267, C166–76.

Falconer, A. E., Carson, R., Johnstone, R., Bird, P., Kehoe, M., and Calvert, J. E. (1993). Distinct IgG1 and IgG3 subclass responses to two streptococcal protein antigens in man: analysis of antibodies to streptolysin O and M protein using standardized subclass-specific enzyme-linked immunosorbent assays. *Immunology*, **79**, 89–94.

Harvey, A. N., Costa, N. D., and Savage, J. R. (1994). Electroporation and streptolysin O – a comparison of poration techniques. *Mutat. Res.*, **315**, 17–23.

Iwamoto, M., Ohno Iwashita, Y. and Ando, S. (1987). Role of the essential thiol group in the thiol-activated cytolysin from Clostridium perfringens. *Eur. J. Biochem.*, **167**, 425–30.

Kehoe, M. A., Miller, L., Walker, J. A., and Boulnois, G. J. (1987). Nucleotide sequence of the streptolysin O (SLO) gene: structural homologies between SLO and other membrane-damaging, thiol-activated toxins. *Infect. Immun.*, **55**, 3228–32.

Momburg, F., Roelse, J., Hammerling, G. J., and Neefjes, J. J. (1994). Peptide size selection by the major histocompatibility complex-encoded peptide transporter. *J. Exp. Med.*, **179**, 1613–23.

Palmer, M., Valeva, A., Kehoe, M., and Bhakdi, S. (1995). Kinetics of streptolysin O self-assembly. *Eur. J. Biochem.*, **231**, 388–95.

Pimplikar, S. W., Ikonen, E., and Simons, K. (1994). Basolateral protein transport in streptolysin O-permeabilized MDCK cells. *J. Cell Biol.*, **125**, 1025–35.

Pinkney, M., Kapur, V., Smith, J., Weller, U., Palmer, M., Glanville, M., et al. (1995). Different forms of streptolysin O produced by *Streptococcus pyogenes* and by *Escherichia coli* expressing recombinant toxin: cleavage by streptococcal cytein protease. *Infect. Immun.*, **63**, 2776–9.

Rapp, S., Soto, U., and Just, W. W. (1993). Import of firefly luciferase into peroxisomes of permeabilized Chinese hamster ovary cells: a model system to study peroxisomal protein import in vitro. *Exp. Cell Res.*, **205**, 59–65.

Smyth, C. J. and Duncan, J. L. (1978). Thiol-activated (oxygen-labile) cytolysins. In *Bacterial toxins and cell membranes* (ed. J. Jeljaszewicz and T. Wadström), pp. 129–83, Academic Press, London.

Tan, A., Bolscher, J., Feltkamp, C., and Ploegh, H. (1992). Retrograde transport from the Golgi region to the endoplasmic reticulum is sensitive to GTP gamma S. *J. Cell Biol.* **116**, 1357–67.

Walev, I., Palmer, M., Valeva, A., Weller, U., and Bhakdi, S. (1995). Binding, oligomerization, and pore formation by streptolysin O in erythrocytes and fibroblast membranes: detection of nonlytic polymers. *Infect. Immun.*, **63**, 1188–94.

Wendland, M. and Subramani, S. (1993). Cytosol-dependent peroxisomal protein import in a permeabilized cell system. *J. Cell Biol.*, **120**, 675–85.

■ S. Bhakdi, M. Kehoe, A. Valeva, U. Weller, M. Palmer: *Institute of Medical Microbiology, University of Mainz, Obere Zahlbacher Str. 67, D 55101 Mainz, Germany*

Cholesterol binding toxins (*Streptococcus, Bacillus, Clostridium, Listeria*)

This family of 50 to 60 kDa pore-forming bacterial toxins (sulfhydryl-activated cytolysins) comprises 19 structurally and antigenically related, chromosomally encoded single-chain soluble proteins endowed with potent lethal and lytic properties resulting from the disruption of the cytoplasmic membrane of eukaryotic cells and that of certain cell organelles. Membrane cholesterol appears as a specific binding site of these toxins and is thought to contribute to their oligomerization in target membranes.

The 19 toxins identified to date are produced by gram-positive aerobic or anaerobic, sporulating or non-sporulating bacteria from the genera *Streptococcus, Bacillus, Clostridium,* and *Listeria* (Smyth and Duncan 1978, Alouf and Geoffroy 1991) listed below in Table 1.

Except for PLY, which is intracytoplasmic, the other toxins are secreted in culture medium during bacterial growth. Among the toxin-producing bacteria, only *Listeria* are intracellular pathogens which grow and release their toxins in host phagocytes or possibly in other cells.

The lethal (cardiotoxic) and cytolytic properties of the toxins are suppressed by oxidants or by thiol-group block-ing agents and restored by reducing agents. The toxins share common epitopes and elicit in humans or in animals neutralizing and precipitating cross-reacting antibodies. The biological properties are irreversibly lost in the presence of very low concentrations of cholesterol and other related 3β-hydroxysterols which interfere with toxin binding on target cells (Alouf and Geoffroy 1991).

A considerable progress in our understanding of structure–activity relationship resulted from the cloning, sequencing, and expression of *tox* genes of 8 out of the 19 toxins of the family.

Table 1

Bacterial genus	Species	Toxin name	Established or suggested[b] toxin and gene acronyms	
Streptococcus	S. pyogenes	Streptolysin 0[a]	SLO	sio
	S. pneumoniae	Pneumolysin[a]	PLY	ply
	S. suis	Suilysin[a]	SUI	sui
Bacillus	B. cereus	Cereolysin 0[a]	CLO	clo
	B. alvei	Alveolysin[a]	ALV	alv
	B. thuringiensis	Thuringiolysin 0[a]	TLO	tlo
	B. laterosporus	Laterosporolysin	LSL	lsi
Clostridium	C. tetani	Tetanolysin*	TTL	ttl
	C. botulinum	Botulinolysin	BTL	btl
	C. perfringens	Perfringolysin 0[a]	PFO	Pfo
	C. septicum	Septicolysin 0	SPL	spl
	C. histolyticum	Histolyticolysin 0	HTL	htl
	C. novyi A (oedematiens)	Novyilysin	NVL	nvl
	C. chauvoei	Chauveolysin	CVL	cvl
	C. bifermentans	Bifermentolysin	BFL	bfl
	C. sordellii	Sordellilysin	SDL	sdl
Listeria	L. monocytogenes	Listeriolysin 0[a]	LLO	llo/lisA
	L. ivanovii	Ivanolysin[a]	ILO	ilo
	L. seeligeri	Seeligerolysin	LSO	lso

Some strains of streptococci of groups C and G also produce streptolysin 0.
[a]Native toxins reported to be purified to apparent homogeneity [b]Author's proposal.

The deduced number of amino acid residues and mol. wt. of mature toxins are the following: SLO (538 AA, 60151 Da; databank accession number M18638; see preceding entry p. 5), PLY (471 AA, 52800 Da, databank accession number: M17717; Walker et al. 1987), PFO (472 AA, 52469 Da; databank accession number: M36704; Tweten 1988b), ALV (469 AA, 51766 Da; databank accession number M62709; Geoffroy et al. 1990), LLO (504 AA, 55842 Da; databank accession number M24199; Mengaud et al. 1988; Domann and Chakraborty 1989), ILO (505 AA, 55961 Da; databank accession number X60461; Haas et al. 1992), and LSO (505 AA, 56371 Da; databank accession number X60462; Haas et al. 1992) Cereolysin gene (clo) has been cloned and expressed in E. coli and B. subtilis (Kreft et al. 1983) but to our knowledge no sequence has been so far reported. A considerable AA sequence homology (stronger at the C-terminal part) was found. It was more pronounced when structurally related AA were taken into account. At the nucleotide sequence level, the homology was lower although detectable, indicating that tox genes have undergone extensive divergence from a common ancestor (Boulnois et al. 1991).

An 11 AA sequence (—ECTGLAWEWWR—) was the longest common motif conserved in each protein (except for single amino acid change in seeligerolysin). It contained the unique Cys residue of the toxins (except for ILO which possesses a second Cys residue beyond the consensus sequence closer to the C-terminus). The thiol group of the Cys residue of the conserved undecapeptide was logically considered as an 'essential' group required for lytic activity as supported by its abrogation by thiol blocking agents. This contention was not supported by site-directed mutagenesis undertaken on PLY, SLO, and LLO (see Michel et al. 1990; Boulnois et al. 1991; Sheehan et al. 1994). Changing the Cys residue to either alanine, serine, or glycine did not affect or only reduced lytic activity suggesting that there is no absolute requirement for the thiol group in the in vitro activity of the toxins. In contrast, the overall structure of the motif appears important (at least in part) for interaction with cholesterol and pore-formation. On the other hand, the use of genetically truncated recombinant PLY showed that the deletion of the six C-terminal AA reduced binding by 96 per cent as also found for PFO (Owen et al. 1994).

The genetic regulation of toxin expression has been particularly investigated for LLO and PFO. The former was shown to be positively regulated by a 27.1 kDa protein encoded by prfa gene the deletion of which besides LLO affects at least four other virulence factors (Schwan et al. 1994; Sheehan et al. 1994). PFO expression was also under positive control of a regulatory gene (pfor) which also affected the expression of clostridal collagenase and hemagglutinin (Shimizu et al. 1994).

■ Purification and sources

The native toxins purified so far are isolated to apparent homogeneity by standard protein chemistry techniques from culture supernatants (except for PLY obtained from disrupted pneumococci) of appropriate toxin-producing strains, grown under culture conditions specific for each bacterial species described in the hereafter cited publications. Most procedures comprise a combination of crude material concentration (ultrafiltration and or salting-out by ammonium sulfate), then ion-exchange column chromatography, followed by gel molecular sieving and (or) hydrophobic (low- or high-pressure) column chromatography. Covalent thiopropyl gel chromatography has been widely used taking advantage of toxin cysteinyl residue. Appropriate purification techniques have been described for PLY (Kanclerski and Möllby 1987; Rubins et al. 1994), PFO (Tweten 1988a), ALV (Geoffroy and Alouf 1983), LLO (Geoffroy et al. 1987; Kreft et al. 1989), ILO (Kreft et al. 1989; Vazquez-Boland et al. 1989), BTL (Haque et al. 1992), and suilysin (Jacobs et al. 1994). The production and purification of fully bioactive recombinant PFO expressed in E.coli (Tweten 1988a) or of PLY expressed in this micoorganism (Mitchell et al. 1989; Rubins et al. 1994) and B.subtilis (Taira et al. 1989) has been reported. The toxins are not commercially available to our knowledge.

The pH optimum for the lytic activity of the various toxins so far studied falls between 6.5 and 7.4 except for LLO with a pH optimum about 5 (no activity at pH 7.0) which may reflect the acid nature of the phagolysosome where it acts (Geoffroy et al. 1987; Sheehan et al. 1994).

■ Toxicity

Microgram quantities of the toxins are lethal for mice, rabbits, and other laboratory animals. The LD_{50} (i.v. route) ranges from 0.2 to 0.8 μg (Smyth and Duncan 1978; Geoffroy et al. 1987).

Hemolytic activity on sheep or rabbit erythrocytes ranges from 0.5 to 2×10^6 hemolytic units/mg of protein (Alouf and Geoffroy 1991). The use of the toxins is safe for operators following normal safety recommendations for bacterial manipulations.

■ Use in cell biology

The lytic properties of the toxins and their specific binding of cholesterol on to eukaryotic cells and organelle membranes have been used for the permeabilization of these structures for the analysis of various cell functions and metabolism, the introduction of exogeneous molecular effectors and the generation of limited lesions in various cells for the isolation of membrane receptors intracellular enzymes or organelles which could not be easily obtained by other methods (see Alouf and Geoffroy 1991; Berthou et al. 1992; Launay et al. 1992; Ahnert-Hilger et al. 1993).

■ References

Ahnert-Hilger, G., Stecher, B., Beyer, C., and Gratzl, M. (1993). Exocytotic membrane fusion as studied in toxin-permealized cells. Methods Enzymol., 221, 139–49.

Alouf, J. E. and Geoffroy, C. (1991). The family of antigenically-related, cholesterol-binding (sulfhydryl-activated) cytolytic toxins. In Sourcebook of bacterial protein toxins (ed. J. E. Alouf and J. H. Freer), pp. 147–86, Academic Press, London.

Berthou, L., Corvaia, N., Geoffroy, C., Mutel, V., Launay, J. M., and Alouf, J. E. (1992). The phosphoinositide pathway of lymphoid cells: labeling after permeabilization by alveolysin, a

bacterial sulfhydryl-activated cytolysin. Eur. J. Cell. Biol., 58, 377–82.

Boulnois, G. J., Paton, J. C., Mitchell, T. J., and Andrew, P. W. (1991). Structure and function of pneumolysin, the multifunctional, thiol-activated toxin of Streptococcus pneumoniae. Mol. Microbiol., 59, 2611–16.

Domann, E. and Chakraborty, T. (1989). Nucleotide sequence of the listeriolysin gene from a Listeria monocytogenes serotype 1/2a strain. Nucleic Acids Res., 17, 6406.

Geoffroy, C. and Alouf, J. E. (1983). Selective purification by thiol–disulfide interchange chromatography of alveolysin, a sulfhydryl-activated toxin of B. alvei. J. Biol. Chem., 255, 9968–72.

Geoffroy, C., Gaillard, J. L., Alouf, J. E., and Berche, P. (1987). Purification, characterization and toxicity of the sulfhydrl-activated hemolysin listeriolysin O from Listeria monocytogenes. Infect. Immun., 55, 1641–46.

Geoffroy, C., Mengaud, J., Alouf, J. E., and Cossart, P. (1990). Alveolysin, the thiol-activated toxin of Bacillus alvei, is homologous to listeriolysin O, perfringolysin O, pneumolysin, and streptolysin O and contains a single cysteine. J. Bacteriol., 172, 7301–5.

Haas, A., Dumbsky, M., and Kreft, J. (1992). Listeriolysin genes : complete sequence of ilo from Listeria ivanovii and of Iso from Listeria seeligeri. Biochim. Biophys. Acta., 1130, 81–4.

Haque, A., Sugimoto, N., Horiguchi, Y., Okabe, T., Miyata, T., Iwanaga, S., et al. (1992). Production, purification, and characterization of botulinolysin, a thiol-activated hemolysin of Clostridium botulinum. Infect. Immun., 60, 71–8.

Jacobs, A. A. C., Loeffen, P. L. W., Van Den Berg, A. J. G., and Storm, P. K. (1994). Identification, purification and characterization of a thiol-activated hemolysin (suilysin) of Streptococcus suis. Infect. Immun., 62, 1742–8.

Kanclerski, K. and Möllby, R. (1987). Production and purification of Streptococcus pneumoniae hemolysin (pneumolysin). J. Clin. Microbiol., 25, 222–5.

Kreft, J., Berger, H., Härtlein, M., Müller, B., Weidinger, G., and Goebel, W. (1983). Cloning and expression in Escherichia coli and Bacillus subtilis of the hemolysin (cereolysin) determinant from Bacillus cereus. J. Bacteriol., 155, 681–9.

Kreft, J., Funke, D., Haas, A., Lottspeich, F., and Goebel, W. (1989). Production, purification and characterization of hemolysins from Listeria ivanovii and Listeria monocytogenes Sv4b. FEMS Microbiol. Lett., 57, 197–202.

Launay, J. M., Geoffroy, C., Mutel, V., Buckle, M., Cesum, A., Alouf, J. E., et al. (1992). One-step purification of the serotonin transporter located at the human platelet plasma membrane. J. Biol. Chem., 267, 11344–51.

Mengaud, J., Vincente, M. F., Chenevert, J., Moniz Pereira, J., Geoffroy, C., Giquel-Sanszey, B., et al. (1988). Expression in Escherichia coli and sequence analysis of the listeriolysin O determinant of Listeria monocytogenes. Infect. Immun., 56, 766–72.

Michel, E., Reich, K. A., Favier, R., Berche, P., and Cossart, P. (1990). Attenuated mutants of the intracellular bacterium Listeria monocytogenes obtained by single amino acid substitutions in listeriolysin O. Mol. Microbiol., 4, 2167–78.

Mitchell, T. J., Walker, J. A., Saunders, F. K., Andrew, P. W., and Boulnois, G. J. (1989). Expression of the pneumolysin gene in Escherichia coli: rapid purification and biological properties. Biochim. Biophys. Acta., 1007, 67–72.

Owen, R. H. G., Boulnois, G. J., Andrew, P. W., and Mitchell, T. J. (1994). A role in cell-binding for the C-terminus of pneumolysin, the thiol-activated toxin of Streptococcus pneumoniae. FEMS Microbiol. Lett., 121, 217–22.

Rubins, J. B., Mitchell, T. J., Andrew, P. W., and Niewoehner, D. E. (1994). Pneumolysin activates phospholipase A in pulmonary artery endothelial cells. Infect. Immun., 62, 3829–36.

Schwan, W. R., Demuth, A., Kuhn, M., and Goebel, W. (1994). Phosphatidylinositol-specific phospholipase C from *Listeria monocytogenes* contributes to intracellular survival and growth of *Listeria innocua. Infect. Immun.*, **62**, 4795–803.

Sheehan, B., Kocks, C., Dramsi, S., Gouin, E., Klarsfeld, A. D., Mengaud, J., *et al.* (1994). Molecular and genetic determinants of the *Listeria monocytogenes* infectious process. *Curr. Top. Microbiol. Immunol.*, **192**, 187–216.

Shimizu, T., Ba-Thein, W., Tamaki, M., and Hayashi, H. (1994). The vir gene, a member of a class of two-component response regulators, regulates the production of perfringolysin O, collagenase, and hemagglutinin in *Clostridium perfringens. J. Bacteriol.*, **176**, 1616–23.

Smyth, C. J. and Duncan, J. J. (1978). Thiol-activated (oxygen-labile) cytolysins. In *Bacterial toxins and cell membranes* (ed. J. Jeljaszewicz and T. Wadström), pp. 129–83, Academic Press, London.

Taira, S., Jalonen, E., Paton, J. C., Sarvas, M., and Runeberg-Nyman, K. (1989). Production of pneumolysin, a pneumococcal toxin, in *Bacillus subtilis. Gene*, **77**, 211–8.

Tweten, R. K. (1988*a*). Cloning and expression in *Escherichia coli* of the perfringolysin O (thetatoxin) gene from *Clostridium perfringens* and characterization of the gene product. *Infect. Immun.*, **565**, 3228–34.

Tweten, R. K. (1988*b*). Nucleotide sequence of the gene for perfringolysin O (theta-toxin) from *Clostridium perfringens*: significant homology with the genes for streptolysin O and pneumolysin. *Infect. Immun.*, **56**, 3235–40

Vazquez-Boland, J.-A., Dominguez, L., Rodriguez-Ferri, E.-F., and Suarez, G. (1989). Purification and characterization of two *Listeria ivanovii* cytolysins, a sphingomyelinase C and a thiol-activated toxin (ivanolysin O). *Infect. Immun.*, **57**, 3928–835.

Walker, J. A., Allen, R. L., Falmagne, P., Johnson, M. K., and Boulnois, G. J. (1987). Molecular cloning, characterization, and complete nucleotide sequence of the gene for pneumolysin, the sulfhydryl-activated toxin of *Streptococcus pneumoniae. Infect. Immun.*, **55**, 1184–9.

■ *Joseph Alouf:*
Institut Pasteur de Lille,
1, rue du Professeur Calmette,
B.P. 245,
F 59019 Lille Cedex,
France

α-Toxin (*Staphylococcus aureus*)

α-Toxin is a 33 kD protein secreted by most pathogenic strains of S. aureus. It binds with both high and low affinity to the membrane of the target cell where it assembles into an oligomeric pore. Pore formation triggers secondary events like eicosanoid production, secretion, endonuclease activation, and cytokine release. Eventually it causes colloid-osmotic lysis of red blood cells, swelling and death of nucleated cells.

Staphylococcal α-toxin is an exotoxin with hemolytic, cytotoxic, dermonecrotic, and lethal activity (Thelestam and Blomqvist 1988; Bhakdi and Tranum-Jensen 1991). It is secreted by most pathogenic strains of *S. aureus* (commonly Wood 46) as a water-soluble polypeptide of 33 kD (293 amino acid residues, sequence accession number to the Swiss Prot databank: P09616 HLA-STAAU; corrections were published (Walker *et al.* 1992)). Its relevance as a virulence factor has been firmly established, at least in animal models. As outlined in Fig. 1, this toxin oligomerizes on the surface of mammalian cells (and liposomes) to form a membrane-embedded oligomer of about 220 kD, which appears in the electron microscope as a hollow cylinder protruding from the plane of the bilayer. It forms crystalline 2D layers on lipid membranes of either natural or artificial origin. From such arrays, a low-resolution three-dimensional map of the oligomer was obtained (Olofsson *et al.* 1988), suggesting it was a hexamer. However, more recent low-resolution X-ray analysis of microcrystals (Gouaux *et al.* 1994) provided evidence that the oligomer is actually an heptamer like the one formed by aerolysin from *Aeromonas hydrophila*.

α-Toxin causes membrane damage to a variety of cells including red blood cells (RBC), platelets, and white cells. RBC damage proceeds in distinct steps: binding to the cell membrane, ion leakage, and, eventually, lysis with release of larger molecules. Nucleated cells can survive its action if adequately protected, e.g. by divalent cations, particularly Zn^{2+} (Bashford *et al.* 1986). Binding occurs either with low affinity to the lipid phase (with a preference for phosphatidylcholine mixed with cholesterol), or with high affinity to a protein receptor present in a few copies on some sensitive cells (e.g rabbit RBC, human platelets, and monocytes, Hildebrand *et al.* 1991). Independent of the binding affinity, damage always requires the assembly of the oligomer. Before inducing cell death, pore formation triggers a number of secondary events: eicosanoid production, secretion, activation of endonucleases, and release of cytokines, which are all explained by a toxin-induced Ca^{2+} influx into cells still possessing an intact cytosolic protein machinery. Early apoptotic effects, at a stage at which the pore is small and permeant only to Na^+, have been detected (Jonas *et al.* 1994). On planar lipid membranes, composed of purified phospholipids, α-toxin forms a water-filled anion-selective pore, with a diameter of around 1 nm (Menestrina 1986). Similar pores were detected in patch-clamped Lettre cells (Korchev *et al.* 1995).

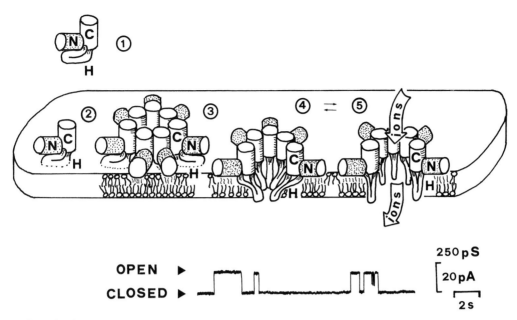

$$\text{OPEN} \blacktriangleright$$

$$\text{CLOSED} \blacktriangleright$$

250 pS
[20 pA]
2 s

Figure 1. Steps leading to the formation of a pore by *Staphylococcus aureus* α-toxin, as resulting from structural, biochemical and, genetic studies.

1. Soluble α-toxin is monomeric and comprises two regions (N-terminal and C-terminal) connected by a glycine-rich flexible hinge (H).

2. The monomer binds in a temperature-independent step to the cell membrane either via a high affinity protein receptor, or, more commonly, via ubiquitous low affinity acceptors (phosphatidylcholine and cholesterol).

3. Absorbed α-toxin monomers oligomerize, via a temperature dependent surface diffusion, to form a non-lytic amphiphilic hexamer.

4. The oligomer can further insert into the lipid matrix of the cell membrane, generating a transmembrane channel precursor which behaves as an integral protein and is resistant to external (or membrane bound) proteases

5. The pore may open allowing the passage of ions and small molecules (up to molecular weight 2000). In artificial membranes the opening of the pore is visualized by an increase of the ionic current (lower trace). The open and closed states of the pore (i.e. states 4 and 5) are in equilibrium and the probability of it being open depends on several factors like pH, divalent cations, and applied voltage.

Table 1 The family of α-toxin and other related leukotoxins from *Staphylococcus aureus*

Toxin	Acronym	Mol. weight (kD)	Identity[a](%)	Similarity[a](%)	Reference[b]
α-toxin	Hla	33	–	–	1
F-component					
α-lysin B	HlgB	34	30.4	12.3	3
leukocidin F	LukF	34	29.4	11.9	2
leukocidin F-R	LukF-R	34	28.0	14.0	2
leukocidin F-PV	LukF-PV	34	26.3	15.7	4
S-component					
α-lysin A	HlgA	32	24.9	11.7	3
α-lysin C	HlgC	32	21.0	14.0	3
leukocidin S	LukS	32	20.6	14.0	2
leukocidin S-R	LukS-R	32	20.6	14.0	2
leukocidin S-PV	LukS-PV	32	21.7	14.0	4

[a] Identity and similarity of the mature forms are referred to α-toxin and were calculated after alignment according to Dayhoff MDM-78 matrix method.
[b] References: 1, Gray and Kehoe 1984; 2, Hunter *et al.* 1993; 3, Cooney *et al.* 1993; 4, Prévost *et al.* 1995.

S. aureus leukocidins and γ-lysins share sequence homology with α-toxin (Cooney *et al.* 1993), suggesting they form a family (Table 1). See also the entry on leukocidins and gamma lysins (p. 94).

■ Purification and sources

α-Toxin purification should be as fast and thorough as possible to avoid subsequent damage by endogenous proteases. Popular procedures use, after ammonium sulfate precipitation, adsorption chromatography on controlled-pore glass followed by ion-exchange (Cassidy and Harshman 1976), or cation-exchange chromatography followed by size-exclusion (Lind *et al.* 1987). α-Toxin can be purchased from Behringwerke (Marburg, Germany) and List.

■ Toxicity

LD_{50} in mice is 50 μg/kg or 1.5 pmoles/kg. Hemolytic activity, on the most sensitive rabbit red blood cells, should be in the range 40 000 HU/mg (tested by addition of 1 volume of 2.5×10^8 red cell/ml for 1 hour at 37 °C). After 4 hours at room temperature it can reach 100 000 HU/mg. Human red blood cells are typically 400 times less sensitive

■ Use in cell biology

α-Toxin is used to selectively permeabilize cells to small molecules (below the cut-off of the pore) while excluding large molecules such as cytoplasmic proteins (above that cut-off). For example, its ability to let Ca^{2+} ions enter the cell while leaving intact the cytoplasmic enzyme cascades and machineries has been exploited to study the minimal requirements for exocytosis (Ahnert-Hilger *et al.* 1985). Since that time a large number of papers reported a similar usage, reviewed by Bhakdi *et al.* (1993). Other applications can be foreseen. For example, in patch-clamp electrophysiology the use of the *perforated patch* as an alternative to the whole-cell configuration is becoming diffuse. In this variant the patch under the pipette is not broken but rather made very permeable by using the channel-forming antibiotic nystatin. Electrical access is gained, but (due to the small molecular weight cut-off of the nystatin pores) internal proteins are retained inside the cell. α-Toxin could be used in place of nystatin providing an alternative choice of pore diameter and selectivity. Relevant for both applications is the possibility to fine tune the size of the α-toxin pores and/or to switch them on and off via different stimuli after direct mutagenesis and modifications of the molecule. Many such possibilities have been reviewed recently (Bayley, 1994) and include opening by a flash of light or by limited proteolysis and closing by Zn^{2+}. The fact that α-toxin self assembles in 2D crystals is potentially relevant for the production of biosensors.

■ References

Ahnert-Hilger, G., Bhakdi, S., and Gratzl, M. (1985). Minimal requirements for exocytosis: a study using PC 12 cells permeabilized with staphylococcal alpha-toxin. *J. Biol. Chem.*, **260**, 12730–4.

Bashford, C. L., Alder, G. M., Menestrina, G., Micklem, K. J., Murphy, J., and Pasternak, C. A. (1986). Membrane damage by haemolytic viruses, toxins, complement and other agents: a common mechanism blocked by divalent cations. *J. Biol. Chem.*, **261**, 9300–8.

Bayley, H. (1994). Triggers and switches in a self-assembling pore-forming protein. *J. Cell. Biochem.*, **56**, 177–82.

Bhakdi, S. and Tranum-Jensen, J. (1991). Alpha-toxin of *Staphylococcus aureus*. *Microbiol. Rev.*, **55**, 733–51.

Bhakdi, S., Weller, U., Walev, I., Martin, E., Jonas, D., and Palmer, M. (1993). A guide to the use of pore-forming toxins for controlled permeabilization of cell membranes. *Med. Microb. Immunol.*, **182**, 167–75.

Cassidy, P. and Harshman, S. (1976). Purification of staphylococcal alpha-toxin by adsorption chromatography on glass. *Infect. Immun.*, **13**, 982–6.

Cooney, J., Kienle, Z., Foster, T. J., and O'Toole, P. W. (1993). The gamma-hemolysin locus of *Staphylococcus aureus* comprises three linked genes, two of which are identical to the genes of the F and S component of leukocidin. *Infect. Immun.*, **61**, 768–71.

Gouaux, J. E., Braha, O., Hobaugh, M. R., Langzhou, S., Cheley, S., Shustak, C., and Bayley, H. (1994). Subunit stoichiometry of staphylococcal α-toxin in crystals and on membranes: a heptameric transmembrane pore. *Proc. Natl. Acad.Sci. USA*, **91**, 12828–31.

Gray, G. S. and Kehoe, M. (1984). Primary sequence of the alpha-toxin gene from *Staphylococcus aureus* Wood 46. *Infect. Immun.*, **46**, 615–18.

Hildebrand, A., Pohl, M., and Bhakdi, S. (1991). *Staphylococcus aureus* α-toxin, dual mechanism of binding to target cells. *J. Biol. Chem.*, **266**, 17195–200.

Hunter, S. E. C., Brown, J. E., Oyston, P. C. F., Sakurai, J., and Titball, R. W. (1993). Molecular genetic analysis of beta-toxin of *Clostridium perfringens* reveals sequence homology with alpha-toxin, gamma toxin, and leukocidin of *Staphylococcus aureus*. *Infect. Immun.*, **61**, 3958–65.

Jonas, D., Walev, I., Berger, T., Liebetrau, M., Palmer, M., and Bhakdi, S. (1994). Novel path to apoptosis: small transmembrane pores created by staphylococcal alpha-toxin in T lymphocytes evoke internucleosomal DNA degradation. *Infect. Immun.*, **62**, 1304–12.

Korchev, Y. E., Alder, G. M., Bakhramov, A., Bashford, C. L., Joomun, B. S., Sviderskaya, E. V., *et al.* (1995). *Staphylococcus aureus* alpha-toxin-induced pores: channel-like behavior in lipid bilayers and clamped cells. *J. Membrane Biol.*, **143**, 143–51.

Lind, I., Ahnert-Hilger, G., Fuchs, G. and Gratzl, M. (1987). Purification of alpha-toxin from *Staphylococcus aureus* and application to cell permeabilization. *Anal. Chem.*, **164**, 84–9.

Menestrina, G. (1986). Ionic channels formed by *Staphylococcus aureus* alpha-toxin: voltage dependent inhibition by di- and trivalent cations. *J. Membrane Biol.*, **90**, 177–90.

Olofsson, A., Kaveus, U., Thelestam, M., and Hebert, H. (1988). The projection structure of α-toxin from *Staphylococcus aureus* in human platelet membranes as analyzed by electron microscopy and image processing. *J. Ultrastruct. Mol. Struct. Res.*, **100**, 194–200.

Prévost, G., Cribier, B., Couppie, P., Petiau, P., Supersac, G., Finck-Barbançon, V., *et al.* (1995). Panton-Valentine leucocidin and gamma-hemolysin from *S. aureus* ATCC 49775 are encoded by distinct genetic loci and have different biological activities. *Infect. Immun.*, **63**, 4121–9.

Thelestam, M. and Blomqvist, L. (1988). Staphylococcal alpha-toxin – recent advances. *Toxicon*, **26**, 51–65.

Walker, B., Krishnasastry, M., Zorn, L., Kasianowicz, J., and Bayley, H. (1992). Functional expression of the α-hemolysin of *Staphylococcus aureus* in intact *Escherichia coli* and in cell lysates. Deletion of five C-terminal amino-acids selectively impairs hemolytic activity. *J. Biol. Chem.*, **267**, 10902–9.

■ *Gianfranco Menestrina and Mercedes Ferreras:*
CNR-ITC Centro Fisica Stati Aggregati,
Via Sommarive 14 ,
38050 Povo ,
Trento,
Italy

δ-Toxin (*Staphylococcus aureus*) and melittin (*Apis mellifera*)

δ-Toxin and melittin (Mel) are 26 residue long peptides, secreted by Staphylococcus aureus *strains and representing about 50 per cent of the dry weight of bee venom, respectively. Both are pleiotropic toxins with no sequence homology, despite this they share most of their biological properties. The first target is the plasma membrane, but the toxins also act on a wide variety of intracellular elements whether membranes, proteins, or nucleic acids. Their behaviour is dominated by their amphipathic character, they are surface active, have rather nonspecific affinity for interfaces, and are classified as 'membrane invading' or 'direct lytic' peptides.*

At first glance the sequence of Mel (Table 1) looks like that of a soap with an apolar segment followed by the strongly basic C_{term} hexapeptide. Careful inspection shows the presence of several polar or less hydrophobic residues, namely $Gly_{1,3}$, Lys_7, $Thr_{10,11}$, Pro_{14}, and Ser_{18} which result in an amphipathic sequence. This is more clearly seen for δ-toxin where comparison of the known sequences from different strains shows a conservative

Table 1 Sequences of the different δ-toxins and melittins and some synthetic binary analogues code: hydrophobic residue; + hydrophilic residue; ~ indifferent. From Fitton *et al.* 1980; McKevitt *et al.* 1990; Habermann 1980; de Grado *et al.* 1982; Blondelle and Houghten 1992; Comut *et al.* 1994

	1				5					10					15					20						26
δ toxin *S. aureus* human	M	A	Q	D	I	I	S	T	I	G	D	L	V	K	W	I	I	D	T	V	N	K	F	T	K	K_{Coo}
S. aureus canine			A								V	E	F			L		A	E			E		I		
S. epidermidis			A																					K	K	K_{Coo}
binary reading	o	o	o/+	+		o	o	+	o/+	o	~	+	o	o	+	o	o	o	+	+	o	+	+	o	o/+	+
melittin *A. mellifera*	G	I	G	A	V	L	K	V	L	T	T	G	L	P	A	L	I	S	W	I	K	R	K	R	Q	QNH_2
A. dorsata						I				S																
A. florea						I				A				T							N			K		
Analogue 1 de Grado *et al.*	L	L	Q	S	L	L	S	L	L	Q	S	L	L	S	L	L	L	Q	W	L	K	R	K	R	Q	ONH_2
binary reading	~	o	~	o	o	o	+	o	o	+	+	o	o	~	o	o	o	+	o	o	+	+	+	+	+	+
L_9K_9 Blondelle *et al.*	AcL	K	L	L	K	K	L	L	K	K	L	K	K	L	L		K	K	LNH_2							
$L_{10}K_5$ Cornut *et al.*	K	L	L	K	L	L	L	K	L	L	L	K	L	L	K											

polar/apolar character at each position (Table 1). Synthetic analogues with a similar periodic polar/apolar alternation, despite massive changes in the residues, are active (de Grado et al. 1982; Alouf et al. 1989). This gives sense to a simplified reading of the sequence in a binary hydrophilic/hydrophobic code. The periodicity all along the sequences nicely fits that of an α-helix, and will result in a typical secondary amphipathic structure. This has proved to be the minimal requirement to get strongly hemolytic peptides, since Leu and Lys copolypeptides properly sequenced in order to generate an amphipathic α-helix (Table 1) mimic the natural toxins (Blondelle and Houghten 1992; Cornut et al. 1994).

In the crystal Mel shows α-helical segments 1 $\rightarrow$ 10 and 14 $\rightarrow$ 26 with an angle of about 120° due to Pro_{14}. Antiparallel helices are packed with all apolar residues buried in the core of a tetramer (Terwilliger and Eisenberg 1982). The structure of δ-toxin is still unsolved, but energy calculations allow the lattice to be filled with antiparallel α-helices, which generate planar oligomers with one polar face, the opposite one being apolar (Raghunathan et al. 1990). When in their aggregated state, in organic solution and when bound to lipids, both toxins are rather similarly α-helical, i.e. about 70 to 80 per cent (Dempsey 1990); for δ-toxin it corresponds to a single helix from Ile_5 to Phe_{23} (Lee et al. 1987; Tappin et al. 1988).

The amphipathicity of the toxins leads to self-association equilibria in buffer and in membrane media. At low concentration, monomeric peptides are almost structureless in buffer but almost totally α-helical in low dielectric organic solvents and membranes. Increasing concentration, ionic strength, or pH stabilizes oligomers in water with a concomitant folding in the α-helix up to 60–70 per cent (Cornut et al. 1993). The hydrophobic effect stabilizes Mel tetramers, favored by phosphate and sulfate. For the δ-toxin, self-association is much more drastic, it occurs in the μM range and association numbers strongly increase up to very large asymmetric aggregates. Out of the hydrophobic burying of the apolar face, self association is increased by intermolecular ion pairing through Asp/Lys residues on the polar face of the α-helices (Thiaudière et al. 1991).

Both toxins interact with lipids and membranes with partition coefficients larger than 10^5. While Mel has a selectivity for negatively charged lipids, δ-toxin has similar affinities whatever the lipid charge (Freer 1986; Cornut et al. 1993). Such a binding to lipids supports the fact that several 10^6 toxins can bind per erythrocyte (Tosteson et al. 1985). Despite a long-running controversy about the location on the membrane, it is now clear that toxins bind essentially parallel to the membrane surface and not as a transmembrane helical rod (Cornut et al. 1996). An increase in the membrane bound concentration of toxins leads to self-association of the peptides and increased perturbation of bilayer structure, whose thickness decreases. This ends up fragmenting lipidic membranes into discoidal or micellar structures (Dufourcq et al. 1986). The length and physical state of the acyl chains, the net charge, and the presence of cholesterol modulate the lytic efficiency on vesicles (Faucon et al. 1995; Monette and Lafleur 1995; Pott and Dufourc 1995). In the presence of an applied potential the two toxins form channels interpreted as barrel–stave aggregates of the amphipathic helices. Increasing association numbers explains the levels of conductance which can vary from a few tens of pS up to several nS for the larger channels (Sansom 1991). The potential-dependent channel activity does not correlate strictly with lysis; one can get channels without lysis and vice versa (Stankowski et al. 1991; Kerr et al. 1995).

■ Biological activities

δ-Toxin and Mel are cytotoxins active on most of the eukaryotic cells. Their hemolytic activity is a colloid osmotic process, i.e. cells first become permeable to small solutes, particularly to ions, and only after swelling hemoglobin does leak (Tosteson et al. 1985). The size of permeants increases differently for the two toxins, on fibroblastes Mel induces faster growing holes compared to δ-toxin (Thelestam and Möllby 1979). When cell lysis occurs, up to 10 per cent of the lipids can be solubilized, which looks like the quantitative solubilization of pure lipid bilayers (Katsu et al. 1989). Mel binds also to proteins facing outside and induces their aggregation (Van Veen and Cherry 1992). Nevertheless, this should play some role; the parallel between lysis of lipid vesicles and cells strongly supports a similar mechanism. The net charge of the toxins is a critical parameter in lysis; the lytic activity and the strong amphipathic and basic character of the peptides are paralleled (Cornut et al. 1993). Cell lysis can be opposed by a variety of compounds such as polyanions, amphipathic proteins like calmoduline, or glycophorin, which either compensate for the osmotic pressure or efficiently bind to the toxins, whether in solution or on the membrane sites (Tosteson et al. 1985; Cornut et al. 1993).

δ-Toxin is not antibacterial (Dhople and Nagaraj 1993), while Mel inhibits the growth of bacteria in the range of 1–10 μM, acting both on gram + and gram –, though in this case it does not lyse E. coli unless the outer membrane has been removed (Katsu et al. 1989). Mel also inhibits the growth of mycoplasma (Cornut et al. 1995). Numerous analogues of Mel, chimeric melittin-cecropin peptides (Cornut et al. 1993) and the simplified and shorter LiKj amphipathic peptides (Blondelle and Houghten 1992; Cornut et al. 1995) proved to be active. Analogues and fragments of δ-toxin with net positive charge proved to become antimicrobial (Dufourcq, Beven, Siffert, Wroblewski in preparation; Dhople and Nagaraj 1995). Then besides an apolar character, the peptides must have a net positive charge in order to be antimicrobial.

Both toxins interfere with numerous enzymatic pathways (Kasimir et al. 1990). Their synergism with phospholipase A_2 (PLA_2) is well documented. Short fragments or analogues of Mel, which are not lytic, increase significantly the PLA_2 activity (Grandbois, Dufourcq and

Salesse, submitted; Hingaro *et al.* 1995). Melittin proved to activate intrinsic PLA_2 in intact cells, mammalian and bacterial ones, and organs, as well as phospholipase C and lipases (Fletcher and Jiang 1993).

■ Purification and sources

δ-Toxin is secreted in the medium by *Staphylococcus aureus* strains at the end of the exponential phase of growth. It can be adsorbed on hydroxylapatite and then eluted at high phosphate concentration. Further purification by HPLC can be performed to get the purest peptide (Birkbeck and Freer 1988; Tappin *et al.* 1988). Synthetic δ-toxin can be used to eliminate any risk of contamination by much more powerful toxins (Alouf *et al.* 1989).

Mel is available from many manufacturers, Serva Feinbiochemicals, Calbiochem, Fluka, or Sigma, it can also be purified from crude bee venom by HPLC. Whatever the origin, special attention should be paid to eliminate PLA_2 contamination, which can lead to severe artefacts especially when using large amounts of toxin compared to the lipid content and/or when looking at long time effects (Dufourcq *et al.* 1984). Synthetic melittin, already available from Bachem, is very helpful in getting rid of such artefacts.

■ Toxicity

LD_{50} of Mel for human erythrocytes (10^7 cells/ml) is 1.2 μg/ml, that of δ-toxin is significantly higher. The lethal dose of Mel for mice is 4 mg/kg (Habermann 1972).

■ Alternative names

δ-Toxin is also referred as δ-haemolysin, δ-hemolysin, or δ-lysin. Melittin spelling is not appropriate since the toxin comes from *Apis mellifera,* but it is too widely used to be changed now.

■ References

Alouf, J. E., Dufourcq, J., Siffert, O., Thiaudière, E., and Geoffroy, C. (1989). Interaction of δ-toxin and synthetic analogues with erythrocytes and lipid vesicles. *Eur. J. Biochem.*, **183**, 381–90.

Birkbeck, T. H. and Freer, J. H. (1988). Purification and assay of staphylococcal δ-lysin. *Meth. Enzymol.*, **165**, 16–23.

Blondelle, S. E. and Houghten, R. A. (1992). Design of model amphipathic peptides having potent antimicrobial activities. *Biochemistry*, **31**, 12688–94.

Cornut, I., Thiaudière, E., and Dufourcq, J. (1993). The amphipathic helix in eytotoxic peptides. In 'The amphipathic helix', (ed. R. M. 'Epand), pp. 173–219, CRC Press, Loc.

Cornut, I., Büttner, K., Dasseux, J. L., and Dufourcq, J. (1994). The amphipathic α-helix concept, application to the *de novo* design of ideally amphipathic Leu, Lys peptides more hemolytic than melittin. *FEBS Lett.*, **349**, 29–33.

Cornut, I., Büttner, K., Beven, L., Duclohier, H., and Dufourcq, J. (1995). Minimal requirements for the design of highly

cytotoxic peptides. In 'Peptides 1994' (ed. H. L. S. Maïa), pp. 666–7, Escom, Leiden.

Cornut, I., Desbat, B., Turlet, Y. M., and Dufourcq, J. (1996). *In situ* study by P structure and orientation of amphipathic peptides at the air-water interface. *Biophys. J.*, **70**, pp. 305–12.

De Grado, W. F., Musso, G. F., Lieber, M., Kaiser E. T. and Kèzdy, F. J. (1982). Kinetics and mechanism of hemolysis induced by melittin and a synthetic anologue. *Biophys. J.*, **37**, 329–38.

Dempsey, C. E. (1990). The action of melittin on membranes. *Biochim. Biophys. Acta*, **1031**, 143–161.

Dhople, V. H. and Nagaraj, R. (1993). δ-Toxin, unlike melittin, has only hemolytic and no antimicrobial activity. *Biosc. Rept.*, **13**, 245–50.

Dhople, V. H. and Nagaraj R. (1995). Analogs having potent automicrobial and hemolytic activities with minimal changes from an inactive 16-residue fragment of δ-toxin. *Prot. Engineer.*, **8**, 315–18.

Dufourcq, J., Dasseux, J. L. and Faucon J. F. (1984). A review of melittin–phospholipid systems. In *Bacterial protein toxins*, pp. 127–38, Academic Press, London.

Dufourcq, J., Faucon, J. F., Fourche, G., Dasseux, J. L., Le Maire, M., and Gulik-Krzywicki, T. (1986). Morphological changes in bilayers induced by melittin: vesicularization, fusion, discoidal particles. *Biochim. Biophys. Acta*, **859**, 33–48.

Faucon, J. F., Bonmatin, J. M., Dufourcq, J., and Dufourc, E. J. (1995). Acyl chain length dependence in the stability of melittin–phosphatidylchloline complexes. *Biochim. Biophys. Acta*, **1234**, 235–43.

Fitton, J. E., Dell, A., and Shaw W. V. (1980). Aminoacid sequence of δ-haemolysin of *Staphylococcus aureus*. *FEBS Lett.*, **115**, 209–12.

Fletcher, J. E. and Jiang, M. S. (1993). Possible mechanisms of action of snake venom cardiotoxins and bee venom melittin. *Toxicon*, **31**, 669–95.

Freer, J. H. (1986). Membrane damage caused by bacterial toxins: recent advances and new challenges. In *Natural toxins* (ed. J. B. Harris), pp. 189–211, Clarendon Press, Oxford.

Habermann, E. (1980). Melittin, structure and activity. In *Natural toxins.* (ed. D. Eaker and T. Wadstrom), pp. 173–81, Pergamon Press, London

Hingaro, I., Pèrez-Payá, E., Pinilla, C., Appel, J. R., Houghten, R. A., and Blondelle, S. E. (1995). Activation of bee venom PLA_2 through a peptide-enzyme complex. *FEBS Lett.*, **372**, 131–4.

Kasimir, S., Schonfeld, W., Alouf, J. E. and Kijning, W. (1990). Effect of *Staphylococcus aureus* δ-toxin on granulocyte and PAF metabolism. *Infect. Immun.*, **58**, 1653–9.

Katsu, T., Kuroko, M., Morikawa, T., Sanchika, K, Fujita, Y., Yamamura, H., *et al.* (1989). Mechanism of membrane damage induced by peptides gramicidin S and melittin. *Biochim. Biophys. Acta.*, **983**, 135–41.

Kerr, I. D., Dufourcq, J., Rice, J. A., Fredkin, D. R. and Sansom, M. S. P. (1995). Ion channel formation by synthetic analogues of δ-toxin. *Biochim. Biophys. Acta*, **1236**, 219–27.

Lee, K. H., Fitton, J. E. and Wüthrich, K. (1987). NMR investigation of the conformation of δ-haemolysin bound to dodecyl phosphocholine. *Biochim. Biophys. Acta*, **911**, 144–53.

McKevitt, A. I., Bjomsonn, G. L., Mauracher, C.A., and Scheifele, D. W. (1990). Amino acid sequence of a delta like toxin from *Staphylococcus epidermidis*. *Infect. Immun.*, **58**, 1473–5.

Monette, M. and Lafleur, M. (1995). Modulation of melittin-induced lysis by surface charge density of membranes. *Biophys. J.*, **68**, 187–95.

Pott, T. and Dufourc, E. J. (1995). Action of Mel on DPPC-cholesterol liquid-ordered phases: a solid state 2H and 31P NHR study. *Biophys. J.*, **68**, 965–77.

Raghunathan, C., Seetharamulu, P., Brooks, B. R., and Guy, H. R. (1990). Models of δ-toxin membrane channels and crystal structures. *Proteins*, **8**, 213–25.

Sansom, M. S. P. (1991). The biophysics of peptide models if ion channels. *Prog. Biophys. Mol. Biol.*, **55**, 139–236.

Stankowski, S., Pawlak, M., Kaïfsheva, E., Robert, C. H., and Schwarz, G. (1991). A combined study of aggregation, membrane affinity and pore activity of natural and modified melittin. *Biochim. Biophys. Acta*, **1069**, 77–86.

Tappin, M. J., Pastore, A., Norton, R. S., Freer, J. H., and Campbell, I. D. (1988). High resolution ¹H NMR study of the structure of δ-hemolysin. *Biochemistry*, **27**, 1643–7.

Terwilliger, T. C. and Eisenberg, D. (1982). The structure of melittin. *J. Biol. Chem.*, **257**, 6016–22.

Thelestam, M. and Möllby, R. (1979). Classification of microbial, plant and animal cytolysins based on their membrane-damaging effects on human fibroblasts. *Biochim. Biophys. Acta*, **556**, 156–69.

Thiaudière, E., Siffert, O., Talbot, J. C., Bolard, J., Alouf, J. E. and Dufourcq, J. (1991). The amphipatic α-helix concept.

Consequences for the structure of δ-toxin in solution and bound to lipids. *Eur. J. Biochem.*, **195**, 203–13.

Tosteson, M. T., Holmes, S. J., Razin, M., and Tosteson, D. C. (1985) Melittin lysis of red cells. *J. Membr. Biol.*, **87**, 35–44.

Van Veen, M. and Cherry, R. J. (1992). The effect of the presence of integral membrane protein in the membrane lytic properties of melittin in reconstituted systems. *FEMS Microbiol. Immun.*, **105**, 147–50.

■ Jean Dufourcq and Sabine Castano:
Centre de Recherche Paul Pascal,
CNRS,
Avenue Schweitzer,
33600 Pessac,
France

Aerolysin (*Aeromonas hydrophila*)

Aerolysin is a 48 kDa channel-forming protein secreted by species in the genus Aeromonas. It binds to a specific receptor on target cells and oligomerizes to form heptamers which can insert into the plasma membrane. This results in the production of well-defined channels that cause disruption of the permeability barrier and cell death.

Aerolysin is secreted as a 52 kDa protoxin by *Aeromonas* sp. (see van der Goot *et al.* 1994, for a recent review). The protein appears to be primarily responsible for the pathogenicity of *A. hydrophila* to mice (Chakraborty *et al.* 1987). The *aerA* structural genes from a number of different species have been sequenced, and the structure of the protein from *A. hydrophila* has been solved (Parker *et al.* 1994). The protein is a dimer in the crystal as well as in solution and each polypeptide chain contains four domains. More than 70 per cent of the molecule is in β-sheet. The protoxin is converted to the active toxin by proteolytic removal of approximately 43 amino acids from the C-terminus (Howard and Buckley 1985; van der Goot *et al.* 1992). This can be accomplished by proteases secreted by the bacteria, as well as by mammalian proteases such as trypsin, chymotrypsin, and furin. Aerolysin binds to a receptor on the surface of target cells. In the case of rats and mice, which are very sensitive to the toxin, the receptor on erythrocytes is a 47 kDa glycoprotein with high affinity (Kd 10^{-9}–10^{-10} M) for aerolysin (Gruber *et al.* 1994). We have recently found that the rat erythrocyte receptor is a member of a family of GPI-anchored membrane proteins that includes the T-cell receptor RT6. On human erythrocytes, which are less sensitive to the toxin, the receptor is likely glycophorin (Garland and Buckley 1988). Binding concentrates aerolysin on the surface of the cell and this promotes oligomerization of the toxin which results in the formation of heptameric structures that can insert into the plasma membrane (Wilmsen *et al.* 1992). The heptamers likely form β-barrels, similar to those produced by bacterial and mitochondria porins. The heptameric channels have some of the properties of porin channels. They are approximately 1.5 nm in diameter (Howard and Buckley 1982), and they are voltage gated (Wilmsen *et al.* 1990).

Aerolysin is a prototype of toxins that likely generate β-barrels in order to form channels. Other examples are the alpha toxins of *Staphylococcus aureus* and *Clostridium septicum*, the oxygen-labile toxins of many Gram-negative species, and *Pseudomonas aeruginosa* hemolysin. These toxins each contain a stretch of amino acids that is similar to the sequence from 250 to 300 in aerolysin. *C. septicum* alpha toxin is homologous throughout its sequence to domains 2, 3, and 4 of aerolysin.

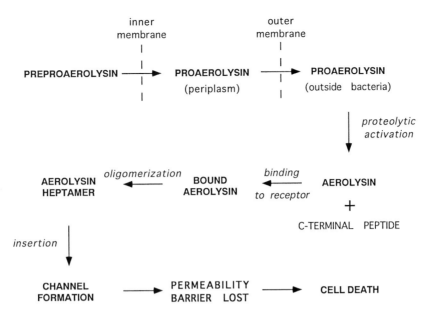

Figure 1. Stages in the formation of a channel by aerolysin. During secretion the protein crosses the inner and outer membranes of the bacteria in separate steps. Proaerolysin can also bind to the receptor on the target cell.

■ Purification and sources

Aerolysin has been isolated from culture supernatants of *Aeromonas hydrophila* (Buckley *et al.* 1981) and *Aeromonas sobria* (Chakraborty *et al.* 1984). Due to the presence of proteases, preparations are often a mixture of the protoxin and the active form and losses can be high, due to oligomerization and aggregation of the toxin. Large amounts of the *A. hydrophila* protoxin can be isolated from culture supernatants of a protease-deficient mutant of *A. salmonicida* containing the cloned *hydrophila* structural gene. Ammonium sulphate precipitation is used, followed by hydroxyapatite and DEAE cellulose chromatography. Proaerolysin can be converted to aerolysin by treatment with trypsin or with trypsin immobilized on agarose beads (Buckley 1990).

■ Toxicity

An LD_{50} for aerolysin has not been established for any species. In one experiment, intravenous injection of 100 ng of toxin killed all the mice tested within 24 hours (Buckley *et al.* 1981). Based on the *in vitro* experiments described above, aerolysin should be less toxic to humans.

■ Uses in cell biology

Aerolysin can be used to permeabilize cells in exactly the same way as *S. aureus* alpha toxin has been used, as well as for the other applications of alpha toxin that have recently been proposed (Krishnasastry *et al.* 1992). It has the advantage that it is more stable than alpha toxin,

and that it can be converted to the active form on demand. Aerolysin has also been used for the isolation of intracellular parasites. This is based on the principle that although the host cells are disrupted by the toxin, parasites like trypanosomes are protected by their glycocalyx. Thus trypanosomes can be easily freed from erythrocytes as a first step in their purification (Pearson *et al.* 1982).

■ References

Buckley, J. T. (1990). Purification of cloned proaerolysin released by a low protease mutant of *Aeromonas salmonicida*. *Biochem. Cell. Biol.*, **68**, 221–4.

Buckley, J. T., Halasa, L. N., Lund, K. D., and MacIntyre, S. (1981). Purification and some properties of the hemolytic toxin aerolysin. *Can. J. Biochem.*, **59**, 430–6.

Chakraborty, T., Schmid, A., Notermans, S., and Benz, R. (1984). Aerolysin of *Aeromonas sobria*: evidence for formation of ion-permeable channels and comparison with alpha-toxin of *Staphylococcus aureus*. *Infect. Immun.*, **58**, 2127–32.

Chakraborty, T., Huhle, B., Hof., H., Bergbauer, H., and Goebel, W. (1987). Marker exchange mutagenesis of the aerolysin determinant in *Aeromonas hydrophila* demonstrates the role of aerolysin in *Aeromonas hydrophila* infections. *Infect. Immun.*, **55**, 2274–80.

Garland, W. J., and Buckley, J. T. (1988). The cytolytic toxin aerolysin must aggregate to disrupt erythrocytes and aggregation is stimulated by human glycoprotein. *Infect. Immun.*, **56**, 1249–54.

Gruber, H. J., Wilmsen, H. U., Cowell, S., Schindler, H., and Buckley, J. T. (1994). Partial purification of the rat erythrocyte receptor for the channel-forming toxin aerolysin and reconstitution into planar lipid bilayers. *Mol. Microbiol.*, **14**, 1093–11.

Howard, S. P., and Buckley, J. T. (1982). Membrane glycoprotein receptor and hole-forming properties of a cytolytic protein toxin. *Biochemistry*, **21**, 1662–7.

Howard, S. P., and Buckley, J. T. (1985). Activation of the hole-forming toxin aerolysin by extracellular processing. *J. Bacteriol.*, **163**, 336–40.

Krishnasastry, M., Walker, B., Zorn, L., Kasianowicz, J., and Bayley, H. (1992). In *Synthetic structures in biological research* (ed. J. M. Schnur and M. Peckerar), pp. 41–51, Plenum Press, New York.

Parker, M. W., Buckley, J. T., Postma, J. P. M., Tucker, A. D., Leonard, K., Pattus, F., *et al.* (1994). Structure of the *Aeromonas* toxin proaerolysin in its water-soluble and membrane-channel states. *Nature*, **367**, 292–5.

Pearson, T. W., Saya, L. E., Howard, S. P., and Buckley, J. T. (1982). The use of aerolysin toxin as an aid for visualization of low numbers of African trypanosomes in whole blood. *Acta Tropica*, **39**, 73–7.

van der Goot, F. G., Lakey, J., Pattus, F., Kay, C. M., Sorokine, O., Van Dorsselaer, A., *et al.* (1992). Spectroscopic study of the activation and oligomerization of the channel-forming toxin aerolysin: identification of the site of proteolytic activation. *Biochemistry*, **31**, 8566–70.

van der Goot, F. G., Pattus, F., Parker, M. W., and Buckley, J. T. (1994). Aerolysin: from the soluble form to the transmembrane channel. *Toxicol.*, **87**, 19–28.

Wilmsen, H. U., Pattus, F., and Buckley, J. T. (1990). Aerolysin, a hemolysin from *Aeromonas hydrophila,* forms voltage-gated channels in planar lipid bilayers. *J. Membr. Biol.*, **115**, 71–81.

Wilmsen, H. U., Leonard, K., Tichelaar, W., Buckley, J. T., and Pattus, F. (1992). The aerolysin membrane channel is formed by heptamerization of the monomer. *EMBO J.*, **11**, 2457–63.

■ *J. Thomas Buckley:*
Department of Biochemistry and Microbiology,
University of Victoria,
Box 3055,
Victoria, BC,
Canada V8W 3P6

Escherichia coli hemolysin

Hemolysin (Hly A) from Escherichia coli *is an extracellular, pore-forming toxin belonging to the RTX family of bacterial cytolysins. It is synthesized as an inactive protoxin of 110 kDa, which is activated within the* E. coli *cell by covalent fatty acid acylation. The secretion of the toxin into the culture medium is accomplished by a specific transport apparatus.* E. coli *hemolysin displays strong cytolytic and cytotoxic activity against a wide range of human and mammalian cells, including erythrocytes and leukocytes, and modulates the functions of several cell types at sublytic concentrations. The toxin represents an important virulence factor in the pathogenesis of extraintestinal* E. coli *infections.*

E. coli hemolysin (Hly A) is the most extensively studied member of the family of RTX toxins, which represents a group of structurally and functionally related cytolysins and cytotoxins produced by a variety of gram-negative bacteria. In addition to *E. coli* hemolysin, the RTX toxins include hemolysins from *Proteus vulgaris* and *Morganella morganii*, hemolysins, and cytotoxins from *Actinobacillus pleuropneumoniae* and *Actinobacillus suis*, leukotoxins from *Pasteurella haemolytica* and *Actinobacillus actinomycetemcomitans*, and the bifunctional adenylate cyclase-hemolysin (adenylate cyclase toxin) from *Bordetella pertussis* (Coote 1992). All RTX toxins are Ca^{2+}-dependent, pore-forming protein toxins. The most pronounced structural feature of these toxins is the presence of a characteristic repetitive domain, which is composed of a toxin-specific number of highly conserved glycine- and aspartate-rich repeat motifs of nine amino acids (XLXGGXGN/DD). The designation RTX (Repeats in ToXins) was deduced from this common structure.

The RTX toxins are synthesized as inactive proteins which have to be activated by an accessory protein prior to secrection from the bacterial cell. In the case of *E. coli* hemolysin, it has been shown that the 110 kDa hemolysin protein (proHlyA) is activated post-translationally by acyl carrier protein-dependent fatty acid acylation of two internal lysine residues, Lys-564 and Lys-690 (Issartel *et al.* 1991; Stanley *et al.* 1994). This covalent modification is mediated by the cytoplasmic protein HlyC (20 kDa) which is coexpressed together with HlyA, but the exact function of HlyC in the activation process is not completely understood.

The extracellular secretion of *E. coli* hemolysin and of the other RTX toxins is accomplished by specific, highly conserved transport systems that consist of at least three-membrane-bound proteins. The transport of *E. coli* hemolysin, in particular, depends on two specific integral proteins of the cytoplasmic membrane, HlyB (80 kDa) and HlyD (55 kDa), and on the common *E. coli* outer membrane protein TolC (52 kDa) (Wagner *et al.* 1983; Wandersman and Delepelaire 1990; Wang *et al.* 1991). These proteins probably form a transenvelope complex that spans both membranes of *E. coli*, allowing the direct secretion of hemolysin into the medium without accumulation in the periplasmic space. Interestingly, HlyB is a member of the superfamily of ATP binding cassette (ABC) transporters and it has been shown that HlyB provides energy for the translocation of HlyA by binding and hydrolyzing ATP (Koronakis *et al.* 1995).

Hemolysin is frequently produced by *E. coli* strains that cause extraintestinal infections in humans, especially those of the urinary tract. *In vivo* studies using several animal models indeed demonstrated that hemolysin significantly contributes to the virulence of these *E. coli* strains (Hacker *et al.* 1983; Cavalieri *et al.* 1984). However, the precise function of *E. coli* hemolysin in the pathogenesis of extraintestinal infections is unclear and probably multifactorial. Due to its cytotoxic activity against leukocytes and other nucleated cell types, it may impair or counteract the host immune defence system and cause tissue damage, thereby promoting the penetration of the bacteria into deeper tissue layers. In addition, the lysis of erythrocytes by hemolysin may stimulate bacterial growth in the host by increasing the level of available iron. Inflammatory mediators, which are released in large amounts from several types of human and mammalian cells upon contact with sublytic doses of *E. coli* hemolysin, may further impair the host and facilitate bacterial spreading.

Alternative names

α-Hemolysin, HlyA

Isolation

E. coli hemolysin was originally identified by its capacity to lyse erythrocytes. The genetic determinant coding for synthesis and secretion of active hemolysin was found to be located either on large plasmids or on the chromosome of hemolytic, wild-type *E. coli* strains. Cloning and sequencing of hemolysin determinants from different strains allowed the further characterization of the toxin.

Gene and sequence

The *E. coli* hemolysin determinant represents the prototype of an RTX toxin determinant. It consists of four structural genes which are arranged in an operon in the order *hly*C, *hly*A, *hly*B, and *hly*D (Felmlee *et al.* 1985; Hess *et al.* 1986) (GenBank accession numbers M10133, M12863, M14107, M81823, X02768, X07565). The gene *hly*A is the structural gene of the *E. coli* hemolysin and encodes the inactive protoxin (proHlyA, 110 kDa, 1024 amino acids); *hly*C encodes the cytoplasmic protein that is necessary for the conversion of proHlyA to the hemolytically active toxin; *hly*B and *hly*D encode the two inner membrane proteins which are required for the secretion of hemolysin (Goebel and Hedgpeth 1982; Wagner *et al.* 1983). The *hly*CABD operon is transcribed from a promoter region located upstream of *hly*C. TolC is encoded by a chromosomal gene not located in the *hly* gene cluster.

Protein

E. coli hemolysin is an extracellularly secreted, fatty acid-acylated protein of 110 kDa which is endowed with hemolytic and cytotoxic activity. The toxin is routinely isolated from the culture supernatant of exponentially growing *E. coli* strains containing the complete hemolysin determinant. Active hemolysin depleted of LPS can be purified from the culture supernatant by precipitation with polyethylene glycol followed by centrifugation in a glycerol density gradient (Jonas *et al.* 1993). Various monoclonal and polyclonal antibodies against *E. coli* hemolysin have been described (Pellett *et al.* 1990; Oropeza-Wekerle *et al.* 1991; Jarchau *et al.* 1994).

Several functional regions have been identified in the hemolysin protein (Fig. 1). The secretion signal necessary for the recognition of HlyA by the transport machinery is located within the C-terminal 50–60 amino acids of HlyA, but the exact structural features of this unprocessed signal are not well understood (Jarchau *et al.* 1994; Kenny *et al.* 1994). Binding of hemolysin to target cell membranes apparently requires the binding of Ca^{2+} to the repeat domain in the C-terminal half of HlyA as well as the HlyC-mediated fatty acid acylation of Lys-564 and Lys-690 in the region proximal to the repeat domain (Ludwig *et al.* 1988; Boehm *et al.* 1990). The transmembrane pore appears to be formed by a domain in the N-terminal half of HlyA which contains putative hydrophobic and amphipathic transmembrane sequences (Ludwig *et al.* 1991). Studies on artificial lipid bilayers have shown that the transmembrane pores generated by *E. coli* hemolysin are unstable, hydrophilic, and cation-specific. From the single channel conductance the pore diameter was estimated to be at least 1 nm (Benz *et al.* 1989).

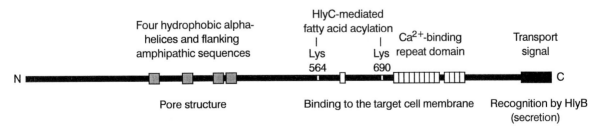

Figure 1. Model of *E. coli* hemolysin (HlyA) showing the location of functional domains.

Biological activities

E. coli hemolysin lyses erythrocytes from a wide range of species and it displays strong cytotoxic activity against a variety of nucleated cells including leukocytes, endothelial cells, and renal epithelial cells (Cavalieri et al. 1984; Jonas et al. 1993). Particularly, granulocytes, monocytes, and human T lymphocytes are killed by nanomolar or even subnanomolar concentrations of the toxin. The cytolytic and cytotoxic activity of E. coli hemolysin is most likely due to the formation of pores in the cytoplasmic membrane of the target cells, which cause a rapid and irreversible depletion of cellular ATP and eventually may lead to osmotic cell lysis. Very low, sublytic concentrations of E. coli hemolysin modulate normal functions of several types of host cells. They induce, for example, the production and release of inflammatory mediators from polymorphonuclear leukocytes, monocytes, and platelets, and cause a contraction of endothelial cells (Grimminger et al. 1991; König et al. 1994). Many of the host cell responses to very low concentrations of E. coli hemolysin are most likely mediated by a defined signal transduction cascade which is triggered by the formation of the transmembrane pores in the target cell membrane.

Biological regulation

Transcription of the hly gene cluster is positively regulated by sequences located upstream of the promoter region, which may act as binding sites for regulatory proteins. In some E. coli strains the expression of hemolysin was found to be negatively regulated by iron in the growth medium.

Mutagenesis studies

The significance of hemolysin as a virulence factor in extraintestinal E. coli infections has been established in several animal models. Experimental infections with hemolytic wild-type E. coli strains and nonhemolytic mutants as well as with genetically engineered, isogenic hemolysin-producing and nonproducing E. coli strains demonstrated that hemolysin is directly involved in the pathogenesis of infection (Hacker et al. 1983; Cavalieri et al. 1984).

Biological interactions

E. coli hemolysin interacts with a variety of eukaryotic cells. However, a specific receptor for hemolysin in target cell membranes has not been identified.

References

Benz, R., Schmid, A., Wagner, W., and Goebel, W., (1989). Pore formation by the Escherichia coli hemolysin: evidence for an association–dissociation equilibrium of the pore-forming aggregates. Infect. Immun., 57, 887–95.

Boehm, D. F., Welch, R. A., and Snyder, I. S. (1990). Domains of Escherichia coli hemolysin (HlyA) involved in binding of calcium and erythrocyte membranes. Infect. Immun., 58, 1959–64.

Cavalieri, S. J., Bohach, G. A., and Snyder, I. S. (1984). Escherichia coli α-hemolysin: characteristics and probable role in pathogenicity. Microbiol. Rev., 48, 326–43.

Coote, J. G. (1992). Structural and functional relationships among the RTX toxin determinants of Gram-negative bacteria. FEMS Microbiol. Rev., 88, 137–62.

Felmlee, T., Pellett, S., and Welch, R. A. (1985). Nucleotide sequence of an Escherichia coli chromosomal hemolysin. J. Bacteriol., 163, 94–105.

Goebel, W. and Hedgpeth, J. (1982). Cloning and functional characterization of the plasmid-encoded hemolysin determinant of Escherichia coli. J. Bacteriol., 151, 1290–8.

Grimminger, F., Scholz, C., Bhakdi, S., and Seeger, W. (1991). Subhemolytic doses of Escherichia coli hemolysin evoke large quantities of lipoxygenase products in human neutrophils. J. Biol. Chem., 266, 14262–9.

Hacker, J., Hughes, C., Hof, H., and Goebel, W. (1983). Cloned hemolysin genes from Escherichia coli that cause urinary tract infection determine different levels of toxicity in mice. Infect. Immun., 42, 57–63.

Hess, J., Wels, W., Vogel, M., and Goebel, W. (1986). Nucleotide sequence of a plasmid-encoded hemolysin determinant and its comparison with a corresponding chromosomal hemolysin sequence. FEMS Microbiol. Lett., 34, 1–11.

Issartel, J.-P., Koronakis, V., and Hughes, C. (1991). Activation of Escherichia coli prohaemolysin to the mature toxin by acyl carrier protein-dependent fatty acylation. Nature, 351, 759–61.

Jarchau, T., Chakraborty, T., Garcia, F., and Goebel, W. (1994). Selection for transport competence of C-terminal polypeptides derived from Escherichia coli hemolysin: the shortest peptide capable of autonomous HlyB/HlyD-dependent secretion comprises the C-terminal 62 amino acids of HlyA. Mol. Gen. Genet., 245, 53–60.

Jonas, D., Schultheis, B., Klas, C., Krammer, P. H., and Bhakdi, S. (1993). Cytocidal effects of Escherichia coli hemolysin on human T lymphocytes. Infect. Immun., 61, 1715–21.

Kenny, B., Chervaux, C., and Holland, I. B. (1994). Evidence that residues –15 to –46 of the haemolysin secretion signal are involved in early steps in secretion, leading to recognition of the translocator. Mol. Microbiol., 11, 99–109.

König, B., Ludwig, A., Goebel, W., and König, W. (1994). Pore formation by the Escherichia coli alpha-hemolysin: role for mediator release from human inflammatory cells. Infect. Immun., 62, 4611–17.

Koronakis, E., Hughes, C., Milisav, I., and Koronakis, V. (1995). Protein exporter function and in vitro ATPase activity are correlated in ABC-domain mutants of HlyB. Mol. Microbiol., 16, 87–96.

Ludwig, A., Jarchau, T., Benz, R., and Goebel, W. (1988). The repeat domain of Escherichia coli haemolysin (HlyA) is responsible for its Ca^{2+}-dependent binding to erythrocytes. Mol. Gen. Genet., 214, 553–61.

Ludwig, A., Schmid, A., Benz, R., and Goebel, W. (1991). Mutations affecting pore formation by haemolysin from Escherichia coli. Mol. Gen. Genet., 226, 198–208.

Ludwig, A., Garcia, F., Bauer, S., Jarchau, T., Benz, R., Hoppe, J., and Goebel, W. (1996). Analysis of the in vivo activation of hemolysin (HlyA) from Escherichia coli. J. Bacteriol., 178, 5422–30.

Oropeza-Wekerle, R. L., Kern, P., Sun, D., Muller, S., Briand, J. P., and Goebel, W. (1991). Characterization of monoclonal antibodies against alpha-hemolysin of *Escherichia coli*. *Infect. Immun.*, **59**, 1846–52.

Pellett, S., Boehm, D. F., Snyder, I. S., Rowe, G., and Welch, R. A. (1990). Characterization of monoclonal antibodies against the *Escherichia coli* hemolysin. *Infect. Immun.*, **58**, 822–7.

Stanley, P., Packman, L. C., Koronakis, V., and Hughes, C. (1994). Fatty acylation of two internal lysine residues required for the toxic activity of *Escherichia coli* hemolysin. *Science*, **266**, 1992–6.

Wagner, W., Vogel, M., and Goebel, W. (1983). Transport of hemolysin across the outer membrane of *Escherichia coli* requires two functions. *J. Bacteriol.*, **154**, 200–10.

Wandersman, C. and Delepelaire, P. (1990). TolC, an *Escherichia coli* outer membrane protein required for hemolysin secretion. *Proc. Natl. Acad. Sci. USA*, **87**, 4776–80.

Wang, R., Seror, S. J., Blight, M., Pratt, J. M., Broome-Smith, J. K., and Holland, I. B. (1991). Analysis of the membrane organization of an *Escherichia coli* protein translocator, HlyB, a member of a large family of prokaryote and eukaryote surface transport proteins. *J. Mol. Biol.*, **217**, 441–54.

■ *Albrecht Ludwig and Werner Goebel:*
  Lehrstuhl für Mikrobiologie,
  Biozentrum der Universität Würzburg,
  Am Hubland,
  D-97074 Würzburg,
  Germany

2

Toxins affecting signal transduction

Introduction

▌Signal transduction

The ability to sense extracellular signals is essential for all living organisms. In animals, the presence of specialized tissues with different functions, that need to coordinate and regulate their activities, makes signal transduction even more essential. Signals are most often delivered to surface receptors by hormones, growth factors, cell to cell contacts, etc. Once triggered by the ligand, the receptor is activated and the signal is transduced across the cell membrane by receptor clustering and/or conformational changes, that results in biochemical changes on the cytosolic side of the membrane. These occur mainly by two mechanisms:

(1) tyrosine kinase receptor phosphorylation of specific sites of its carboxyterminal cytoplasmic domain, that recruits SH2-transducers to start a cascade of phosphorylation of intracellular targets;

(2) modification of a receptor-coupled G protein that transduces the signal to enzymes, releasing second messengers such as cyclic-AMP (cAMP), inositol-triphosphate (ITP), diacyglycerol (DG), to a cytoplasmic kinase etc.

So far, no toxins have been described that interfere with signal transduction mediated by receptors with a cytoplasmic kinase domain. In marked contrast, many toxins are known to interfere with G protein mediated signal transduction, or with one of the steps downstream from a G protein regulated process. As shown in Fig. 1, these are pertussis toxin, cholera and *Escherichia coli* LT toxins, exoenzyme C3 of *Clostridium botulinum*, the *Bordetella pertussis* adenylate cyclase-hemolysin, and the edema factor of *Bacillus antracis,* which are described in this chapter, and also exoenzyme S of *Pseudomonas aeruginosa* and the *C. botulinum* toxins A and B, which are described in other chapters of this manual.

▌G proteins

G proteins are a family of GTP-binding proteins composed of three subunits (α, β and γ), that are involved in signal transduction across the membrane in animals, plants, and fungi (Hepler and Gilman 1992; Linder and Gilman 1992). The α subunit is usually anchored to the cell membrane by a 14-carbon fatty acid (myristic acid) (Jones *et al.* 1990; Mumby *et al.* 1990; Spiegel *et al.* 1991).

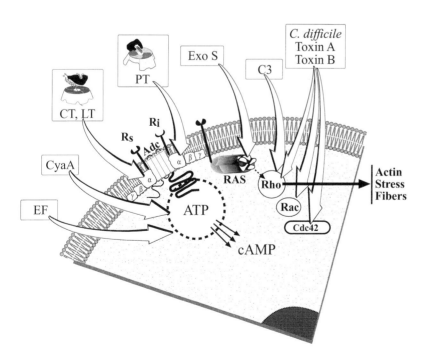

Figure 1. Schematic representation showing the bacterial toxins affecting signal transduction and their targets in an eukaryotic cell.

Typically, a signal that is sensed by a receptor on the surface of eukaryotic cells is received by the α subunit of the G protein, which consequently binds GTP, dissociates from the β and γ subunits, and transmits the signal to the enzymes that release second messengers such as adenylate cyclase, phospholipase C, and cyclic GMP phosphodiesterase. Adenylate cyclase, shown as an example in Fig. 1, is regulated by two classes of receptors that can transmit their signals to two different GTP-binding proteins: G_s and G_i. G_s receives signals from the stimulatory receptors and activates the adenylate cyclase, whereas G_i receives signals from the inhibitory receptors and inhibits the activity of adenylate cyclase. In addition to G_s and G_i, the family of G proteins contains several other proteins, including G_t, G_o, G_{olf} (Jones and Reed 1989; Hepler and Gilman 1992), and several other G proteins with a similar function, that have not yet been fully characterized. As shown in Fig. 2, cholera toxin ADP-ribosylates Arg-201 of

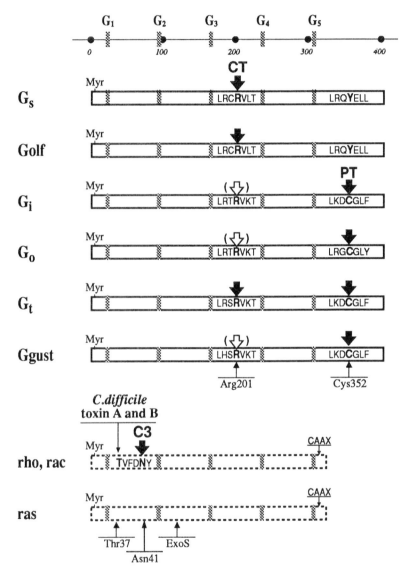

Figure 2. Schematic representation of the proteins involved in signal transduction that are modified by bacterial protein toxins. The top line shows the numbers of the amino acids. G_1 to G_5 represent the regions of homology within the target G proteins. Myr and CAAX indicate sites that are modified by the addition of myristic acid or a farnesyl residue, respectively. Amino acids are indicated in the one-letter code; those that are ADP-ribosylated by the toxins are in bold letters and indicated by large arrows. Small arrows indicate the glucosylation of Thr37 by *C. difficile* toxins A and B. G_{gust} stands for Gustducin. CT, PT, and C3 stand for cholera, pertussis, and *Clostridium botulinum* C3 toxins, respectively.

the α subunit of G_s, G_t and G_{olf} (Van Dop et al. 1984), whereas it ADP-ribosylates the corresponding Arg in G_i and G_o only when these are receptor-activated (Iiri et al. 1989). Pertussis toxin ADP-ribosylates Cys-352 residue in G_i, G_o, G_t, G_{gust}, and other G proteins, but is not able to modify G_s and G_{olf}, which have a tyrosine in position 352 (Katada and Ui 1982; Bokoch et al. 1983; West et al. 1985; Fong et al. 1988). ADP-ribosylation of G_s by cholera or LT toxins causes an irreversible block of the GTPase activity of the G_s protein, leading to the constitutive activation of adenylate cyclase, which results in intracellular accumulation of the second messenger cAMP. ADP-ribosylation of G_i by pertussis toxin uncouples this protein from the receptor in such a way that it is unable to transmit inhibitory signals to the adenylate cyclase.

■ Ras, Rho, Rac, CDC 42, and the small GTP-binding proteins

Ras, Rho, Rac, CDC 42, and other small GTP-binding proteins belong to a large family of regulatory proteins that are known to be involved in controlling diverse essential cellular functions, including growth, differentiation, cytoskeletal organization, intracellular vesicle transport, and secretion. They usually contain a carboxy terminal CAAX box that is modified by addition of a farnesyl lipid moiety, increasing the hydrophobicity and localizing them to the plasma membrane or to the membranes of intracellular vesicles (Hall 1990). The oncoprotein Ras, which is localized to the plasma membrane, controls cell growth and differentiation (Egan et al. 1993), and is a target of ADP-ribosylation by *Pseudomonas* exoenzyme S (Coburn et al. 1989). Following ADP-ribosylation, the GTPase activity of Ras is decreased (Tsai et al. 1985). Rho and Rac are structurally similar proteins involved in the organization of actin (Ridley and Hall 1992; Ridley et al. 1992), formation of focal adhesions, membrane ruffling, etc. They are ADP-ribosylated on Asn-41 by *C. botulinum* C3 (Aktories et al. 1989; Sekine et al. 1989), and glucosylated on Thr-37 by *C. difficile* toxins A and B (Aktories and Just 1995). Both modifications of Rho cause actin depolymerization and cell rounding.

■ Invasive adenylate cyclases

Some of the toxins are enzymes that bypass membrane signal transduction and interfere with signal transduction by generating directly second messengers, such as cAMP. Two of these enzymes, the adenylate cyclase-hemolysin of *B. pertussis* and the edema factor of *B. antracis*, are totally unrelated proteins that have evolved different ways to reach the intracellular space. However, once inside the cell they display the same catalytic activity: they use ATP to produce cAMP, thus altering the normal levels of this second messenger and producing the same effects that are indirectly induced by cholera and LT toxins.

■ References

Aktories, K. and Just, I. (1995). Monoglucosylation of low-molecular-mass GTP-binding Rho proteins by clostridial cytotoxins. *Trends Cell Biol.*, **5**, 441–3.

Aktories, K., Braun, U., Rosener, S., Just, I., and Hall, A. (1989). The rho gene product expressed in *E. coli* is a substrate of botulinum ADP-ribosyltransferase C3. *Biochem. Biophys. Res. Commun.*, **158**, 209–13.

Bokoch, G. M., Katada, T., Northup, J. K., Hewlett, E. L., and Gilman, A. G. (1983). Identification of the predominant substrate for ADP-ribosylation by islet activating protein. *J. Biol. Chem.*, **258**, 2072–5.

Coburn, J., Wyatt, R. T., Iglewski, B. H., and Gill, D. M. (1989). Several GTP-binding proteins, including p21c-H-ras, are preferred substrates of *Pseudomonas aeruginosa* exoenzyme S. *J. Biol. Chem.*, **264**, 9004–8.

Egan, S. E., Giddings, B. W., Brooks, M. W., Buday, L., Sizeland, A. M., and Weinberg, R. A. (1993). Association of Sos Ras exchange protein with Grb2 is implicated in tyrosine kinase signal transduction and transformation. *Nature*, **363**, 45–51.

Fong, H. K., Yoshimoto, K. K., Eversole-Cire, P., and Simon, M. I. (1988). Identification of a GTP-binding protein by alpha subunit that lacks an apparent ADP-ribosylation site for pertussis toxin. *Proc. Natl. Acad. Sci. USA*, **85**, 3066–70.

Hall, A. (1990). The cellular functions of small GTP-binding proteins. *Science*, **249**, 635–40.

Hepler, J. R. and Gilman, A. G. (1992). G proteins. *Trends Biochem. Sci.*, **17**, 383–7.

Iiri, T., Tohkin, M., Morishima, N., Ohoko, Y., Ui, M., and Katada, T. (1989). Chemotactic peptide receptor-supported ADP-ribosylation of a pertussis toxin substrate GTP-binding protein by cholera toxin in neutrophil-type HL-60 cells. *J. Biol. Chem.*, **264**, 21394–400.

Jones, D. T. and Reed, R. R. (1989). G_{olf}: an olfactory neuron specific-G protein involved in odorant signal transduction. *Science*, **244**, 790–5.

Jones, T. L., Simonds, W. F., Merendino, J. J., Jr, Brann, M. R., and Spiegel, A. M. (1990). Myristoylation of an inhibitory GTP-binding protein alpha subunit is essential for its membrane attachment. *Proc. Natl. Acad. Sci. USA*, **87**, 568–72.

Katada, T. and Ui, M. (1982). ADP-ribosylation of the specific membrane protein of C6 cells by islet-activating protein associated with modification of adenylate cyclase activity. *J. Biol. Chem.*, **257**, 7210–16.

Linder, M. E. and Gilman, A. G. (1992). G proteins. *Sci. Am.*, **July**, 56–65.

Mumby, S. M., Heukeroth, R. O., Gordon, J. I., and Gilman, A. G. (1990). G-protein alpha-subunit expression, myristoylation, and membrane association in COS cells. *Proc. Natl. Acad. Sci. USA*, **87**, 728–32.

Ridley, A. J. and Hall, A. (1992). The small GTP-binding protein rho regulates the assembly of focal adhesions and actin stress fibers in response to growth factors. *Cell*, **70**, 389–99.

Ridley, A. J., Paterson, H. F., Johnston, C. L., Diekmann, D., and Hall, A. (1992). The small GTP-binding protein rac regulates growth factor-induced membrane ruffling. *Cell*, **70**, 401–10.

Sekine, A., Fujiwara, M., and Narumuya, S. (1989). Asparagine residue in the rho gene product is the modification site for botulinum ADP-ribosyltransferase. *J. Biol. Chem.*, **264**, 8602–5.

Spiegel, A. M., Backlund, P. S., Jr., Butrynski, J. E., Jones, T. L., and Simonds, W. F. (1991). The G protein connection: molecular basis of membrane association. *Trends Biochem. Sci.*, **16**, 338–41.

Tsai, S. C., Adamik, R., Moss, J., Vaughan, M., Manne, V., and Kung, H.F. (1985). Effects of phospholipids and

ADP-ribosylation on GTP hydrolysis by *Escherichia coli*-synthesized Ha-ras-encoded p21. *Proc. Natl. Acad. Sci. USA*, **82**, 8310–14.

Van Dop, C., Tsubokawa, M., Bourne, H. R., and Ramachandran, J. (1984). Amino acid sequence of retinal transducin at the site ADP-ribosylated by cholera toxin. *J. Biol. Chem.*, **259**, 696–8.

West, R. E., Jr, Moss, J., Vaughan, M., Liu, T., and Liu, T. Y. (1985). Pertussis toxin-catalyzed ADP-ribosylation of transducin. Cysteine 347 is the ADP-ribose acceptor site. *J. Biol. Chem.*, **260**, 14428–30.

■ *Rino Rappuoli:*
IRIS, Chiron Vaccines Immunobiological Research Institute in Siena,
Via Fiorentina 1,
53100 Siena,
Italy

Cholera toxin (*Vibrio cholerae*)

Cholera toxin (CT) is a protein of 82 kDa secreted by Vibrio cholerae, *that is responsible for the watery diarrhea typical of cholera, a severe disease causing 150 000 deaths each year. The toxicity is due to the ADP-ribosylating activity of the A subunit, an enzyme which modifies the α-subunit of G_s, a GTP binding protein involved in signal transduction that regulates the production of cAMP in the eukaryotic cells. This induces an alteration in the ion transport, with an increase in chloride secretion and an inhibition in sodium absorption, causing the diarrhea characteristic of the disease.*

CT is composed of two subunits, A and B, of 27 and 57.5 kDa respectively, held together by noncovalent interactions. The B subunit which contains the receptor binding site, is formed by five identical monomers, each of 11.5 kDa which assemble into a ring like oligomeric structure having a central hole. The CT receptor is the glycosfingolipid, GM1 ganglioside. The A subunit is composed of the enzymatically active A1 domain that is connected to the B subunit through the A2 domain, a long alpha helix inserted into the central hole of the B subunit (Fig. 1) (Rappuoli and Pizza 1991; Sixma *et al.* 1991a; Sixma *et al.* 1993; Merritt *et al.* 1994).

The genes coding for the two subunits are located in a 1.5 Kb DNA region (accession numbers: X58786 for El Tor and X58785 for Classical strains, respectively), and are transcribed from the same promoter that is regulated by ToxR (Mekalanos *et al.* 1983; Miller and Mekalanos 1984). The A and B subunits are synthesized as precursors with a signal peptide of 18 and 21 aminoacids, respectively, which are cleaved during the translocation to the cytoplasm. Holotoxin assembly takes place in the periplasm (Hirst *et al.* 1984; Hirst and Holmgren 1987). Only the fully assembled AB_5 toxin or the B_5 pentamer are secreted into the culture supernatant by a specialized secretory apparatus, that is present only in *Vibrio cholerae* strains (Sandkvist *et al.* 1993). Activation of the A subunit requires proteolitic cleavage of the loop which links the A1 to the A2 domains and reduction of a disulphide bridge that holds the two domains together (Lai *et al.* 1981). *V. cholerae* secretes a specific protease that cleaves the above loop (Booth *et al.* 1984). The A subunit binds NAD and transfers the ADP-ribose group to an Arg residue present in an... LRX**R**VXT motif located in the central part of all G proteins (Gill and Woolkalis 1991). However, only G_s, G_t, and G_{olf} are substrates for CT. G_i, G_o and G_{gust}, under physiological conditions, are not modified by CT. *In vitro*, the ADP-ribosylation activity of CT is enhanced by ADP-ribosylation factors (ARFs), which are a family of small GTP-binding proteins (Moss and Vaughan 1991).

Toxins with a similar A–B structure and with the same mechanism of action are diphtheria toxin (DT), exotoxin A of *Pseudomonas aeuriginosa* (PAETA), pertussis toxin (PT), and heat-labile enterotoxin (LT) (Domenighini *et al.* 1991, 1994). LT shares with CT an identical AB_5 structure and an 80% amino acid homology (Dallas and Falkow 1980; Domenighini *et al.* 1995).

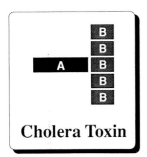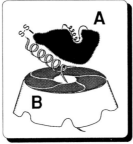

Figure 1. A schematic representation of the subunit composition (left panel), and of the 3D structure (right panel) of cholera toxin.

Structure and mutants

The 3D structure of LT has been solved (Sixma *et al.* 1991*b*, 1993). The A subunit shows a strong homology of the cavity containing the active site with the other ADP-ribosylating toxins. This cavity is formed by a β-strand bent over an α-helix, which form the floor and the ceiling of the cavity, respectively. Inside this cavity, many residues important for the enzymatic activity, have been identified (Domenighini *et al.* 1991, 1994). Based on this information, many non-toxic CT mutants have been generated by site-directed mutagenesis of residues located in the catalytic site (Hase *et al.* 1994; Fontana *et al.* 1995). The best characterized non toxic mutant of CT is CT-K63, containing the Ser → Lys sub-stitution in position 63 (Fontana *et al.* 1995). Mutants in the B subunit with different GM1 binding affinity have been constructed. One of them, unable to bind the receptor, containing a Gly33 → Asp has been crystallized and the structure has been determined (Shoham *et al.* 1995).

Purification and sources

Cholera toxin can be purified from the supernatant of *V. cholerae* strain by precipitation with Sodium hexa-metaphosphate at pH 4.5 (Mekalanos 1988), and chrom-atography on CM-Sepharose column (Fontana *et al.* 1995). Cholera toxin can be purchased from Sigma Chemical Company, St. Louis, Mo., USA. The mutants in the A and the B subunits are available from the authors.

Toxicity

LD_{50} in NIH Swiss mice after intraperitoneal injection is 33.3 μg ± 7.3 μg (Dragunsky *et al.* 1992). However, a basic technique to study the toxic response to CT is the rabbit ileal loop. In this assay, 50 ng of CT are able to induce fluid accumulation in the intestine of rabbit. Studies of toxicity of CT in humans have shown that the ingestion of 25 μg of CT induces over 20 litres of diarrhea, while 5 μg of CT induces 1 to 6 litres of diarrhea (Levine *et al.* 1983).

Use in cell biology

CT has been widely used for the dissection of G_s-mediated signal transduction and the consequent regulation of adenylate cyclase activity in animal and even in plant cells (Beffa *et al.* 1995). In addition to this classical use, today it can be used to study tissue permeability to ions (Ussing chambers) (Field *et al.* 1969), and to study the structure–function of ARF proteins. *In vivo*, CT has been shown to be an immunogen of unusually high efficiency (Elson and Ealding 1984), and to have a potent adjuvant activity at the mucosal level (Holmgren *et al.* 1993). These properties may be used to dissect the mechanisms of antigen uptake and presentation *in vitro*. The presence of a KDEL sequence at the carboxy-terminal end of the A subunit suggests that this protein also has a mechanism of reaching the Golgi, and that the protein may be also used to study intracellular trafficking of endocytic vesicles and proteins (Lencer *et al.* 1995).

References

Beffa, R., Szell, M., Meuwly, P., Pay, A., Vögeli-Lange, R., Métraux, J.-P., *et al.* (1995). Cholera toxin elevates pathogen resistance and induces pathogenesis-related gene expression in tobacco. *EMBO J.*, **14(23)**, 5753–61.

Booth, B. A., Boesman-Finkelstein, M., and Finkelstein, R. A. (1984). *Vibrio cholerae* hemagglutinin/protease nicks cholera enterotoxin. *Infect. Immun.*, **45**, 558–60.

Dallas, W. S. and Falkow, S. (1980). Amino acid homology between cholera toxin and *Escherichia coli* heat labile toxin. *Nature*, **288**, 499–501.

Domenighini, M., Montecucco, C., Ripka, W. C., and Rappuoli, R. (1991). Computer modelling of the NAD binding site of ADP-ribosylating toxins: active-site structure and mechanism of NAD binding. *Mol. Microbiol.*, **5**, 23–31.

Domenighini, M., Magagnoli, C., Pizza, M., and Rappuoli, R. (1994). Common features of the NAD-binding and catalytic site of ADP-ribosylating toxins. *Mol. Microbiol.*, **14 (1)**, 41–50.

Domenighini, M., Pizza, M., Jobling, M. G., Holmes, R. K., and Rappuoli, R. (1995). Identification of errors among database sequence entries and comparison of correct amino acid sequences for the heat-labile enterotoxins of *Escherichia coli* and *Vibrio cholerae*. *Mol. Microbiol.*, **15(6)**, 1165–67.

Dragunsky, E. M., Rivera, E., Aaronson, W., Dolgaya, T. M., Hochstein, H. D., Habig, W. H., *et al.* (1992). Experimental evaluation of antitoxic protective effect of new cholera vaccines in mice. *Vaccine*, **10**, 735–6.

Elson, C. O. and Ealding, W. (1984). Generalized systemic and mucosal immunity in mice after mucosal stimulation and cholera toxin. *J. Immunol.*, **132**, 2736–41.

Field, M., Fromm, D., Wallace, C. K., and Greenough III, W.B. (1969). Stimulation of active chloride secretion in small intes-tine by cholera exotoxin. *J. Clin. Invest.*, **48**, 24a.

Fontana, M. R., Manetti, R., Giannelli, V., Magagnoli, C., Marchini, A., Domenighini, M., *et al.* (1995). Construction of non toxic derivatives of cholera toxin and characterization of the immunological response against the A subunit. *Infect. Immun.*, **63**, 2356–60.

Gill, D. M . and Woolkalis, M. J. (1991). Cholera toxin-catalyzed [32P]ADP-ribosylation of proteins. *Methods Enzymol.*, **195**, 267–80.

Hase, C. C., Thai, L. S., Boesmanfinkelstein, M., Mar, V. L., Burnette, W. N., Kaslow, H. R., *et al.* (1994). Construction and characterization of recombinant *Vibrio cholerae* strains producing inactive cholera toxin analogs. *Infect. Immun.*, **62**, 3051–7.

Hirst, T. R. and Holmgren, J. (1987). Transient entry of enterotoxin subunits into the periplasm occurs during their secretion from *Vibrio cholerae*. *J. Bacteriol.*, **169**, 1037–45.

Hirst, T.R., Sanchez, J., Kaper, J.B., Hardy, S.J. and Holmgren, J. (1984). Mechanism of toxin secretion by *Vibrio cholerae* investigated in strains harboring plasmids that encode heat-labile enterotoxins of *Escherichia coli*. *Proc. Natl. Acad. Sci. USA*, **81**, 7752–6.

Holmgren, J., Lycke, N. and Czerkinsky, C. (1993). Cholera Toxin and cholera-B subunit as oral mucosal adjuvant and antigen vector systems. *Vaccine*, **11**, 1179–84.

Lai, C., Cancedda, F., and Duffy, L. K. (1981). ADP-ribosyltransferase activity of cholera toxin polypeptide A1 and the effect of limited trypsinolysis. *Biochem. Biophys. Res. Commun.*, **102**, 1021–7.

Lencer, W. I., Constable, C., Moe, S., Jobling, M. G., Webb, H. M., Ruston, S., *et al.* (1995). Targeting of cholera toxin and *Escherichia coli* heat labile toxin polarized epithelia: role of COOH-terminal KDEL. *J. Cell Biol.*, **131(4)**, 951–62.

Levine, M. M., Kaper, J. B., Black, R. E. and Clements, M. L. (1983). New knowledge on pathogenesis of bacterial infections as applied to vaccine development. *Microbiol. Rev.*, **47**, 510–50.

Mekalanos, J. (1988). Production and purification of cholera toxin. *Methods Enzymol.*, **165**, 169–75.

Mekalanos, J. J., Swartz, D. J., Pearson, G. D., Harford, N., Groyne, F. and de Wilde, M. (1983). Cholera toxin genes: nucleotide sequence, deletion analysis and vaccine development. *Nature*, **306**, 551–7.

Merritt, E. A., Sarfaty, S., Vandenakker, F., Lhoir, C., Martial, J. A., and Hol, W. G. J. (1994). Crystal structure of cholera toxin B-pentamer bound to receptor G(M1) pentasaccharide. *Protein Sci.*, **3**, 166–75.

Miller, M. and Mekalanos, J. J. (1984). Synthesis of cholera toxin is positively regulated at the transcriptional level by toxR. *Proc. Natl. Acad. Sci. USA*, **81**, 3471–5.

Moss, J. and Vaughan, M. (1991). Activation of cholera toxin and *Escherichia coli* heat-labile enterotoxins by ADP-ribosylation factors, a family of 20 kDa guanine nucleotide-binding proteins. *Mol. Microbiol.*, **5**, 2621–7.

Rappuoli, R. and Pizza, M. (1991). Structure and evolutionary aspects of ADP-ribosylating toxins. In *Sourcebook of bacterial protein toxins* (ed. J. Alouf and J. Freer), pp. 1–20, Academic Press, London and San Diego.

Sandkvist, M., Morales, V. and Bagdasarian, M. (1993). A protein required for secretion of cholera toxin through the outer membrane of *Vibrio cholerae*. *Gene*, **123**, 81–6.

Shoham, M., Scherf, T., Anglister, J., Levitt, M., Merritt, E. A., and Hol, W. G. J. (1995). Structural diversity in a conserved cholera toxin epitope involved in ganglioside binding. *Protein Sci.*, **4**, 841–8.

Sixma, T. K., Pronk, S. E., Kalk, K. H., Wartna, E. S., van Zanten, B. A., Witholt, B., *et al.* (1991a). Crystal structure of a cholera toxin-related heat-labile enterotoxin from *E. coli* [see comments]. *Nature*, **351**, 371–7.

Sixma, T. K., Pronk, S. E., Kalk, K. H., Wartna, E. S., van Zanten, B. A., Witholt, B., *et al.* (1991b). Crystal structure of a cholera toxin-related heat-labile enterotoxin from E. coli. *Nature*, **351**, 371–7.

Sixma, T. K., Kalk, K. H., Vanzanten, B. A. M., Dauter, Z., Kingma, J., Witholt, B., *et al.* (1993). Refined Structure of *Escherichia coli* heat-labile enterotoxin, a close relative of cholera toxin. *J. Mol. Biol.*, **230**, 890–918.

■ *Mariagrazia Pizza, Maria Rita Fontana, and Rino Rappuoli:*
IRIS, Chiron Vaccines Immunobiological Research Institute in Siena,
Via Fiorentina 1,
53100 Siena,
Italy

Heat-labile enterotoxins (*Escherichia coli*)

E. coli *heat-labile enterotoxins (LT) are classified into serogroups LT-I and LT-II. Each holotoxin structurally resembles cholera toxin (CT) and consists of one A polypeptide and five identical B polypeptides. The B pentamer binds to cell surface gangliosides and triggers uptake of holotoxin by receptor-mediated endocytosis. After proteolytic cleavage and reduction of polypeptide A, fragment A1 enters the cytoplasm, activates adenylate cyclase by ADP-ribosylation of the regulatory protein $G_s\alpha$, and initiates a cascade of signal transduction pathways that mediate the biologic effects of LT. In the small intestine, LT stimulates secretion of fluid and electrolytes into the lumen, resulting in watery diarrhea.*

The structure, function, and role in disease of the heat-labile enterotoxins (LT) of *Escherichia coli* have been recently reviewed (Hirst 1995; Hol *et al.* 1995; Holmes *et al.* 1995). Enterotoxigenic *E. coli* (ETEC) that cause acute diarrhea in humans and animals produce plasmid-encoded LT-I, heat-stable toxin (ST), or both. ETEC from pigs and humans produce different antigenic variants of LT-I, respectively called LTp-I and LTh-I. LT-II-producing ETEC are isolated from animals and food but rarely from humans, and they have not yet been shown to cause diarrheal disease. Two antigenic variants of LT-II, designated LT-IIa and LT-IIb have been characterized. A comparative analysis of database entries for the amino acid sequences of the A and B polypeptides of the recognized LT-I, LT-II, and CT variants, including correction of known errors in the published sequences, was recently reported

(Domenighini *et al.* 1995). Original sources for the nucleotide sequences of genes that encode these toxins are cited in the same publication.

The structures of LTp-I, CT, the B pentamer of CT, and co-crystals of some of these proteins with galactose, lactose or the oligosaccharide from ganglioside GM1 have been reported (Sixma *et al.* 1993; Merritt *et al.* 1994a, b; Zhang *et al.* 1995a, b), and the structure of LT-IIb has recently been solved (van den Akker *et al.* 1996). The folds adopted by the homologous polypeptides of the type I and type II enterotoxins are very similar (Fig. 1), despite striking differences in the primary sequences among the B polypeptides. In each of these toxins the five B polypeptides associate to form a planar ring with a central pore that is penetrated by and interacts with the carboxy-terminal A2 domain of polypeptide A. The A1

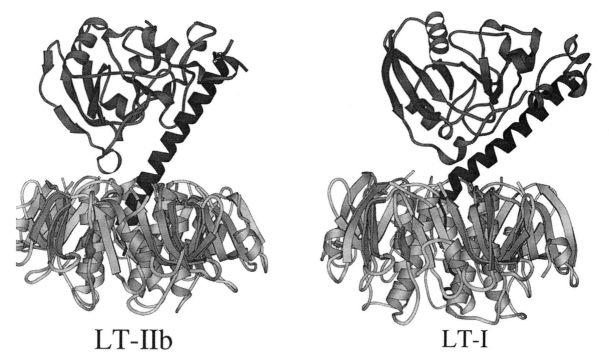

LT-IIb   LT-I

Figure 1. Structure of *Escherichia coli* heat-labile enterotoxins LT-IIb and LT-1. The structure of LT-IIb is remarkably similar to that of LT-I, both in the A subunit and in the B subunits. There is a 24° difference in the relative orientation between the A and B subunits in the two holotoxins, and in LT-IIb the A2 helix is longer and extends further into the pore of the B pentamer than in LT-I. Genetic data indicate that the locations of the ganglioside GD1a-binding sites in LT-IIb are similar to those demonstrated for the ganglioside GM1-binding sites in LT-1. (Figure kindly provided by Focco van den Akker and Wim G. J. Hol.)

domain, which corresponds to an inactive conformation of the ADP ribosyltransferase enzyme, extends outward from the plane of the B pentamer and is located on the opposite side from the ganglioside-binding domains of the B polypeptides. The B pentamers of LTp-I and CT have five binding sites for the GM1 oligosaccharide, and each GM1 oligosaccharide interacts predominantly with residues in a single B polypeptide.

The A and B polypeptides for representative heat-labile enterotoxins from *E. coli* and *V. cholerae* are compared in Table 1. The A polypeptides of type I and type II enterotoxins exhibit the greatest degree of sequence homology within their A1 domains (Pickett *et al.* 1987, 1989;

Domenighini *et al.* 1995; Holmes *et al.* 1995). The A polypeptides of the type I enterotoxins CT, LTh-I, and LTp-I are related more closely to one another than to the type II enterotoxins LT-IIa and LT-IIb, and vice versa. Conserved features of the A polypeptides include: NAD:$G_s\alpha$ ADP-ribosyltransferase activity of fragment A1 that is stimulated by the ARF family of GTP-binding proteins; high susceptibility to trypsin of the surface-exposed, disulfide-linked oligopeptide loop between domains A1 and A2 that is selectively cleaved during generation of fragments A1 and A2 ; the ability of the A2 domain to interact with complementary B polypeptides during assembly of holotoxin; and the presence of KDEL

Table 1(a) Comparison of characteristics of A polypeptides of heat-labile enterotoxins

Toxin	Number of amino acid residues	Location of cysteine residues	Carboxyl terminal sequence	ADP ribosylation of		Stimulation by ARF	Percentage amino acid homology with		
				$G_s\alpha$	agmatine		CT	LTh-I	LT-IIa
CT	240	187, 199	KDEL	active	active	yes	–	81.3	50.4
LTh-I	240	187, 199	RNEL	active	active	yes	81.3	–	54.2
LTp-I	240	187, 199	RDEL	active	active	yes	81.3	99.2	54.2
LT-IIa	241	185, 197	KDEL	active	<1%	yes	50.4	54.2	–
LT-IIb	243	185, 197	KDEL	active	<1%	yes	51.7	53.4	75.5

or a related endoplasmic retention sequence at the carboxyl terminus of A2 that affects intracellular trafficking of the enterotoxins in target cells (Lee *et al.* 1991; Holmes *et al.* 1995; Lencer *et al.* 1995). A structural model for conversion of the inactive A1 domain of LTp-I to enzymatically active fragment A1 was recently proposed (van den Akker *et al.* 1995), and the structure of pseudomonas exotoxin A domain III complexed with AMP and nicotinamide (Li *et al.* 1995) has provided new insights into the probable mode of binding of NAD by other ADP ribosylating toxins, including the heat-labile enterotoxins. The B polypeptides of the type I enterotoxins are highly homologous, but they show little primary sequence identity with the B polypeptides of the type II enterotoxins (Domenighini *et al.* 1995; Holmes *et al.* 1995). The B polypeptides of LT-IIa and LT-IIb are homologous, but the extent of identity between them is less than that observed among the B polypeptides of the type I enterotoxins. These differences in amino acid sequence correlate with differences in ganglioside-binding activities among the members of the heat-labile enterotoxin family (Fukuta *et al.* 1988).

Purification and sources

The heat-labile enterotoxins are produced as periplasmic proteins by *E. coli*. Typically they are produced by overexpression of the cloned operons, and the recombinant enterotoxins are purified from lysed cells. CT and LT-I are easily purified in high yield by affinity chromatography on immobilized D-galactose, followed by elution with D-galactose (Uesaka *et al.* 1994). LT-IIa and LT-IIb are purified by combinations of chromatographic methods adapted for the characteristics of each toxin (Guth *et al.* 1986; Holmes *et al.* 1986).

Toxicity

The heat-labile enterotoxins alter the functions of target cells but do not kill them; they are cytotonic rather than cytotoxic. Treatment with trypsin of enterotoxins that have an intact ('un-nicked') A polypeptide increases their toxicity by 10- to 100-fold against cultured target cells that do not activate the enterotoxins via endogenous proteases. The most sensitive tissue culture systems (mouse Y1 adrenal tumor cells or Chinese hamster ovary cells) detect activated heat-labile enterotoxins with minimum detectable doses of approximately 1–25 picograms, and in Y1 adrenal cell culture systems the specific toxicity of LT-II is significantly greater than that of LT-I or CT (Guth *et al.* 1986; Holmes *et al.* 1986). The parenteral lethality of LT-I is assumed to be comparable with that of CT (about 250 μg i.v. for mice), since their activities are similar in several nonlethal assays, but smaller doses of LT-I or CT are lethal for susceptible animals by the enteral route (Gill 1982). A 25 μg oral dose of CT in volunteers caused diarrhea comparable to that of cholera (Levine *et al.* 1983). Unlike CT and LT-I, LT-II does not stimulate secretion in rabbit ileal segment assays, and data concerning lethality of LT-II in animals are not available (Guth *et al.* 1986; Holmes *et al.* 1986).

Uses in cell biology

CT and LT-I are widely used in cell biology as probes for cyclic AMP and G protein-dependent signal transduction pathways (Harnett, 1994). CT and LT-I are potent mucosal immunogens with a variety of immunomodulatory activities including musosal adjuvanticity (Xu-Amano *et al.* 1994). They are studied extensively as probes for the immune system and as tools for vaccine development . CT and LT-I enter target cells by receptor-mediated endocytosis and undergo both retrograde transport into Golgi cisternae and endoplasmic reticulum as well as transcytosis in polarized epithelial cell cultures (Lencer *et al.* 1995). They are under active investigation as probes for endocytotic and transcytotic pathways in eukaryotic cells. Uptake and retrograde transport of CT-B by neurons is widely used in neurobiology for mapping of neural pathways (Datiche *et al.* 1995). The recently characterized LT-II enterotoxins have not yet been used extensively in cell biology. Because they differ from CT and LT-I in ganglioside-binding specificity (Table 1(b)), they may be particularly useful in comparative studies with CT and LT-I to investigate the role of various gangliosides in endocytosis, transcytosis, and signal transduction pathways.

Acknowledgment

Research in the author's laboratory on heat-labile enterotoxins of *Escherichia coli* was supported in part by Public

Table 1(b) Comparison of characteristics of B polypeptides of heat-labile enterotoxins

Toxin	Number of amino acid residues	Location of cysteine residues	Relative binding activity for gangliosides	Percentage amino acid homology with		
				CT	LTh-I	LT-IIa
CT	103	9, 86	GM1 >GD1b	–	81.6	11.7
LTh-I	103	9, 86	GM1 >GD1b >GM2	81.6	–	10.7
LTp-I	103	9, 86	Not determined	80.6	97.1	10.7
LT-IIa	100	10, 81	GD1b >GD1a,GT1b,GQ1b GM1,GD2 >GM2,GM3	11.7	10.7	–
LT-IIb	99	10, 81	GD1a >GT1b >GM3	13.6	13.6	56.4

Health Service Grant AI14107 from the National Institute of Allergy and Infectious Diseases.

■ References

Datiche, F., Luppi, P. H., and Cattarelli, M. (1995). Serotonergic and non-serotonergic projections from the raphe nuclei to the piriform cortex in the rat: a cholera toxin B subunit (CTb) and 5-HT immunohistochemical study. *Brain Res.*, **671**, 27–37.

Domenighini, M., Pizza, M., Jobling, M. G., Holmes, R. K., and Rappuoli, R. (1995). Identification of errors among database sequence entries and comparison of correct amino acid sequences for the heat-labile enterotoxins of *Escherichia coli* and *Vibrio cholerae* [letter]. *Mol. Microbiology*, **15**, 1165–7.

Fukuta, S., Magnani, J. L., Twiddy, E. M., Holmes, R. K., and Ginsburg, V. (1988). Comparison of the carbohydrate-binding specificities of cholera toxin and *Escherichia coli* heat-labile enterotoxins LTh-I, LT-IIa, and LT-IIb. *Infect. Immun.*, **56**, 1748–53.

Gill, D. M. (1982). Bacterial toxins: a table of lethal amounts. *Microbiol. Rev.*, **46**, 86–94.

Guth, B. E., Twiddy, E. M., Trabulsi, L. R., and Holmes, R. K. (1986). Variation in chemical properties and antigenic determinants among type II heat-labile enterotoxins of *Escherichia coli*. *Infect. Immun.*, **54**, 529–36.

Harnett, M. M. (1994). Analysis of G-proteins regulating signal transduction pathways. *Meth. in Mol. Biol.*, **27**, 199–211.

Hirst, T. R. (1995). Biogenesis of cholera toxin and related oligomeric enterotoxins. In *Bacterial toxins and virulence factors in disease* (ed. J. Moss, B. Iglewski, M. Vaughan, and A. T. Tu), pp. 123–84, Marcel Dekker, New York.

Hol, W. G. J., Sixma, T. K., and Merritt, E. A. (1995). Structure and function of E. coli heat-labile enterotoxin and cholera toxin B pentamer. In *Bacterial toxins and virulence factors in disease* (ed. J. Moss, B. Iglewski, M. Vaughan, and A. T. Tu), pp. 185–223, Marcel Dekker, New York.

Holmes, R. K., Twiddy, E. M., and Pickett, C. L. (1986). Purification and characterization of type II heat-labile enterotoxin of *Escherichia coli*. *Infect. Immun.*, **53**, 464–73.

Holmes, R. K., Jobling, M. G., and Connell, T. D. (1995). Cholera toxin and related enterotoxins of gram negative bacteria. In *Bacterial toxins and virulence factors in disease* (ed. J. Moss, B. Iglewski, M. Vaughan, and A. T. Tu), pp. 225–55, Marcel Dekker, New York.

Lee, C. M., Chang, P. P., Tsai, S. C., Adamik, R., Price, S. R., Kunz, B. C., *et al.* (1991). Activation of *Escherichia coli* heat-labile enterotoxins by native and recombinant adenosine diphosphate-ribosylation factors, 20-kD guanine nucleotide-binding proteins. *J. Clin. Invest.*, **87**, 1780–6.

Lencer, W. I., Constable, C., Moe, S., Jobling, M. G., Webb, H. M., Ruston, S., *et al.* (1995). Targeting of cholera toxin and *Escherichia coli* heat labile toxin in polarized epithelia: role of COOH-terminal KDEL. *J. Cell. Biol.*, **131**, 951–62.

Levine, M. M., Kaper, J.B , Black, R. E., and Clements, M. L. (1983). New knowledge on pathogenesis of bacterial enteric infections as applied to vaccine development. *Microbiol. Rev.*, **47**, 510–50.

Li, M., Dyda, F., Benhar, I., Pastan, I., and Davies, D. R. (1995). The crystal structure of *Pseudomonas aeruginosa* exotoxin domain III with nicotinamide and AMP: conformational differences with the intact exotoxin. *Proc. Natl. Acad. Sci. USA*, **92**, 9308–12.

Merritt, E. A., Sarfaty, S., van den Akker, F., L'hoir, C., Martial, J. A. and Hol, W. G. J. (1994a). Crystal structure of cholera toxin B-pentamer bound to receptor GM1 pentasaccharide. *Protein. Sci.*, **3**, 166–75.

Merritt, E. A., Sixma, T. K., Kalk, K. H., van Zanten, B. A. M., and Hol, W. G. J. (1994b). Galactose binding site in *E. coli* heat-labile enterotoxin (LT) and cholera toxin (CT). *Mol. Microbiol.*, **13**, 745–53.

Pickett, C. L., Weinstein, D. L., and Holmes, R. K. (1987). Genetics of type IIa heat-labile enterotoxin of *Escherichia coli*: operon fusions, nucleotide sequence, and hybridization studies. *J. Bacteriol.*, **169**, 5180–7.

Pickett, C. L., Twiddy, E. M., Coker, C., and Holmes, R. K. (1989). Cloning, nucleotide sequence, and hybridization studies of the type IIb heat-labile enterotoxin gene of *Escherichia coli*. *J. Bacteriol.*, **171**, 4945–52.

Sixma, T. K., Kalk, K. H., van Zanten, B. A., Dauter, Z., Kingma, J., Witholt, B. *et al.* (1993). Refined structure of *Escherichia coli* heat-labile enterotoxin, a close relative of cholera toxin. *J. Mol. Biol.*, **230**, 890–918.

Uesaka, Y., Otsuka, Y., Lin, Z., Yamasaki, S., Yamaoka, J., Kurazono, H., and Takeda, Y. (1994). Simple method of purification of *Escherichia coli* heat-labile enterotoxin and cholera toxin using immobilized galactose. *Microb. Pathog.*, **16**, 71–6.

van den Akker, F., Merritt, E.A., Pizza, M., Domenighini, M., Rappuoli, R. and Hol, W.G. (1995). The Arg7Lys mutant of heat-labile enterotoxin exhibits great flexibility of active site loop 47-56 of the A subunit. *Biochemistry*, **34**, 10996–1004.

van den Akker, F., Sarfaty, S., Twiddy, E. M., Connell, T. D., Holmes, R. K., and Hol, W. G. J. (1996). Crystal structure of a new heat-labile enterotoxin, LT-IIb. *Structure*, **4**, 665–78.

Xu-Amano, J., Jackson, R. J., Fujihashi, K., Kiyono, H., Staats, H. F., and McGhee, J. R. (1994). Helper Th1 and Th2 cell responses following mucosal or systemic immunization with cholera toxin. *Vaccine*, **12**, 903–11.

Zhang, R. G., Scott, D. L., Westbrook, M. L., Nance, S., Spangler, B. D., and Shipley, G. G. W., EM. (1995a). The three-dimensional crystal structure of cholera toxin. *J. Mol. Biol.*, **251**, 563–73.

Zhang, R. G., Westbrook, M. L., Westbrook, E. M., Scott, D. L., Otwinowski, Z., Maulik, P. R., *et al.* (1995b). The 2.4 A crystal structure of cholera toxin B subunit pentamer: choleragenoid. *J. Mol. Biol.*, **251**, 550–62.

■ *Randall K. Holmes:*
Department of Microbiology,
Campus Box B-175,
University of Colorado Health Sciences Center,
4200 East Ninth Avenue,
Denver, CO 80262,
USA

Pertussis toxin (*Bordetella pertussis*)

Pertussis toxin (PT), a bacterial protein toxin secreted in the culture supernatant by Bordetella pertussis, *intoxicates eukaryotic cells by ADP-ribosylating the alpha subunit of G proteins, thus interfering with receptor mediated signal transduction. PT which is a major virulence factor of* B. pertussis *and the main component of anti-pertussis vaccines, has been instrumental in the discovery of G proteins and the characterization of signal transduction in eukaryotic cells.*

PT (Sekura *et al.* 1985) is a protein of 105 000 daltons composed of five noncovalently linked subunits named S1 through S5, and organized into two functional domains called A and B (Tamura *et al.* 1982). The A domain, which is composed of the S1 subunit, is an enzyme that intoxicates eukaryotic cells by ADP-ribosylating their GTP-binding proteins (Rappuoli and Pizza 1991). The enzyme binds NAD and transfers the ADP-ribose group to a cysteine residue present in an ...X**C**GLX motif, located at the carboxy-terminal region of the alpha subunit of many G proteins such as G_i, G_o, G_t, G_{gust}, and others. G_s and G_{olf}, which in this position have a tyrosine instead of the cysteine, are not substrates for PT (Domenighini *et al.* 1995).

The B domain is a nontoxic oligomer formed by subunits S2, S3, S4, and S5 which are present in a 1:1:2:1 ratio. This domain binds the toxin receptor on the surface of eukaryotic cells and facilitates the translocation of the S1 subunit across the cellular membrane, so that it can reach the target proteins. The B oligomer binds to glycoproteins having a branched mannose core with *N*-acetyl glucosamine attached, such as fetuin (Sekura and Zhang 1985). The crystal structure of the protein is known (Stein *et al.* 1994). The genes coding for the five subunits of PT are located in 3.2 kilobases of DNA, and are organized in a bacterial operon, in the following order: S1, S2, S4, S5, S3. Each gene contains a sequence coding for a signal peptide (Locht and Keith 1986; Nicosia *et al.* 1986). The five subunits are cotranslationally secreted into the bacterial periplasm where the protein is assembled into the holotoxin. Secretion of the toxin occurs only after assembly of the holotoxin, or of the B domain (Pizza *et al.* 1990), and requires the product of the nine-gene *ptl* operon, which is located downstream from the PT operon and is regulated by the same promoter (Covacci and Rappuoli 1993; Weiss *et al.* 1993).

■ Mutants

Many mutants containing single or multiple amino acid substitutions in the A or B subunits have been constructed by site-directed mutagenesis of the PT gene. The most popular one is PT-9K/129G, a nontoxic mutant, where Arg9 and Glu129 have been replaced by Lys and Gly respectively (Pizza *et al.* 1989). The mutation in these two amino acids which are located in the catalytic site make the S1 subunit enzymatically inactive and the mutant PT completely nontoxic. The nontoxic mutant is being used for vaccination against pertussis and therefore is produced in gram quantities (Rappuoli *et al.* 1995). Although not commercially available, it can be obtained from the authors of this entry.

■ Purification and sources

PT can be purified from the culture supernatant of *B. pertussis* by a two step procedure involving affi-gel blue chromatography, followed by affinity chromatography on a fetuin-containing column (Sekura *et al.* 1983). PT is also commercially available available from several companies, including Sigma Chemical Company, St. Louis, Mo., USA.

■ Toxicity

PT is lethal at 0.5 μg/mouse. However, PT induces a number of toxic effects at a very low concentration. It renders mice susceptible to death following histamine challenge at 0.5 ng/mouse, potentiates anaphylaxis at 9.5 ng/mouse, enhances insulin secretion at 2 ng/mouse, induces lymphocytosis at 8 ng/mouse, it enhances the IgE production at 0.1 ng/mouse (Munoz 1985), and it changes the permeability of intestinal tissue at a few nanograms per rat. Some of these effects can be measured in animals

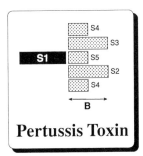
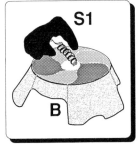

Figure 1. A schematic representation of the subunit composition (left panel), and of the 3D structure (right panel) of pertussis toxin.

a long time after the toxin administration. For instance, the change in intestinal permeability is still present eight months after toxin treatment (Kosecka *et al.* 1994).

■ Use in cell biology

PT was instrumental in the discovery of G_i protein and has been extensively used for the dissection of receptor mediated signal transduction (Katada and Ui 1982; Bokoch *et al.* 1983). Generally, receptors have been divided into 'pertussis toxin sensitive' and 'pertussis toxin insensitive', depending on whether PT interfered or not with signal transduction. The nontoxic mutants are the best controls to understand whether the observed effects are due to the modification of the G proteins mediated by the enzymatic activity of the toxin, or just to the binding of the toxin to its receptor.

■ References

Bokoch, G. M., Katada, T., Northup, J. K., Hewlett, E. L., and Gilman, A. G. (1983). Identification of the predominant substrate for ADP-ribosylation by islet activating protein. *J. Biol. Chem.*, **258**, 2072–5.

Covacci, A. and Rappuoli, R. (1993). Pertussis toxin export requires accessory genes located downstream from the pertussis toxin operon. *Mol. Microbiol.*, **8**, 429–34.

Domenighini, M., Pizza, M., and Rappuoli, R. (1995). Bacterial ADP-ribosyltransferases. In *Bacterial toxins and virulence factors in disease* (ed. J. Moss, B. Iglewski, M. Vaughan, and A. T. Tu), pp. 59–80, Marcel Dekker, New York.

Katada, T. and Ui, M. (1982). ADP ribosylation of the specific membrane protein of C6 cells by islet-activating protein associated with modification of adenylate cyclase activity. *J. Biol. Chem.*, **257**, 7210–16.

Kosecka, U., Marshall, J. S., Crowe, S. E., Bienenstock, J., and Perdue, M. H. (1994). Pertussis toxin stimulates hypersensitivity and enhances nerve-mediated antigen uptake in rat intestine. *Am. J. Physiol-Gastrointest. L.*, **30**, G745–53.

Locht, C. and Keith, J. M. (1986). Pertussis toxin gene: nucleotide sequence and genetic organization. *Science*, **232**, 1258–64.

Munoz, J. J. (1985). Biological activity of pertussigen (pertussis toxin). In *pertussis toxin* (ed. R. D. Sekura, J. Moss, and M. Vaughan), pp. 1–18, Academic Press, Orlando, Florida.

Nicosia, A., Perugini, M., Franzini, C., Casagli, M. C., Borri, M. G., Antoni, G., *et al.* (1986). Cloning and sequencing of the pertussis toxin genes: operon structure and gene duplication. *Proc. Natl. Acad. Sci. USA*, **83**, 4631–5.

Pizza, M., Covacci, A., Bartoloni, A., Perugini, M., Nencioni, L., de Magistris, M. T., *et al.* (1989). Mutants of pertussis toxin suitable for vaccine development. *Science*, **246**, 497–500.

Pizza, M., Covacci, A., Bugnoli, M., Manetti, R., and Rappuoli, R. (1990). The S1 subunit is important for pertussis toxin secretion. *J. Biol. Chem.*, **265**, 17759–63.

Rappuoli, R. and Pizza, M. (1991). Structure and evolutionary aspects of ADP-ribosylating toxins. In *Sourcebook of bacterial protein toxins* (ed. J. Alouf and J. Freer), pp. 1–20, Academic Press, London and San Diego.

Rappuoli, R., Douce, G., Dougan, G., and Pizza, M. (1995). Genetic detoxification of bacterial toxins: a new approach to vaccine development. *Int. Arch. Allergy Immunol.*, **108**, 327–33.

Sekura, R. D. and Zhang, Y. (1985). Pertussis toxin: Structural elements involved in the interaction with cells. In *Pertussis toxin* (ed. R. D. Sekura, J. Moss, and M. Vaughan), pp. 45–64, Academic Press, Orlando, Florida.

Sekura, R. D., Fish, F., Manclark, C. R., Meade, B., and Zhang, Y. L. (1983). Pertussis toxin. Affinity purification of a new ADP-ribosyltransferase. *J. Biol. Chem.*, **258**, 14647–51.

Sekura, R., Moss, J., and Vaughan, M. (1985). *Pertussis toxin*, Academic Press, Orlando, Florida; pp. V–255.

Stein, P. E., Boodhoo, A., Armstrong, G. D., Cockle, S. A., Klein, M. H., and Read, R. J. (1994). The crystal structure of pertussis toxin. *Structure*, **2**, 45–57.

Tamura, M., Nogimori, K., Murai, S., Yajima, M., Ito, K., Katada, T., *et al.* (1982). Subunit structure of the islet-activating protein, pertussis toxin, in conformity with the A-B model. *Biochemistry*, **21**, 5516–22.

Weiss, A. A., Johnson, F. D., and Burns, D. L. (1993). Molecular characterization of an operon required for pertussis toxin secretion. *Proc. Natl. Acad. Sci. USA*, **90**, 2970–4.

■ *Rino Rappuoli and Mariagrazia Pizza:*
IRIS, Chiron Vaccines Immunobiological Research
Institute in Siena,
Via Fiorentina 1,
53100 Siena,
Italy

Exoenzyme C3 (*Clostridium botulinum*)

Exoenzyme C3 (C3) is an M$_r$ 23.6 kDa protein produced by strains C and D of C. botulinum. C3 is an ADP-ribosyltransferase which inactivates selectively the small GTP-binding proteins Rho (A, B, and C) by modification of its asparagine 41. C3 is not a toxin and therefore cannot enter into cells. Upon introduction into cells, for instance by microinjection, C3 inactivation of Rho is followed by the collapse of actin stress fibres. C3 is used in cell biology to inactivate Rho 'in vitro' and 'in vivo'.

ADP-ribosylating toxins have proved to be valuable tools for studying their target proteins. The targets generally suffer important changes in function upon modification, and they become labelled when radioactive NAD is used. In this respect C3 has proved to be extremely useful in understanding the physiological role of the small GTP-binding protein Rho (Narumiya and Morii 1993; Hall 1994).

C3 is an M_r 23.6 kDa ADP-ribosyl transferase which modifies selectively the small GTP-binding proteins Rho (Chardin *et al.* 1989; Aktories *et al.* 1989) on Asparagine 41 (Sekine *et al.* 1989), a residue in close proximity of the Rho effector domain. C3 is secreted by the C and D types of *C. botulinum* (Aktories *et al.* 1987; Rubin *et al.* 1988). Several C3 related exoenzymes have been isolated from other bacteria than *C. botulinum*, such as certain strains of *Staphylococcus aureus* (EDIN) (Inoue *et al.* 1991), *Clostridium limosum* (Just *et al.* 1992a), *Bacillus cereus* (Just *et al.* 1992b) and *Legionella pneumophila* (Belyi *et al.* 1991). However, C3 from *C. botulinum* is the most utilized Rho specific ADP-ribosyltransferase to inactivate and label this GTP-binding protein *in vitro* and *in vivo*. Other members of the Rho family of small GTP-binding proteins such as Rac and Cdc42 poor substrates for C3 (100-folds less for Rac, 400-folds less for Cdc42). C3 has apparently no role as a bacterial virulence factor and an extremely low toxicity upon injection into mice (about 400 μg kg^{-1} ip).

C3 is composed of 211 amino acids (the preprotein has an M_r of 7.8 kDa and 251 amino acids) and is strongly positively charged (pI = 10.4). C3 is a very stable protein and renatures easily after heat- or detergent-denaturation. The C3 gene is borne on a mobile element by *C. botulinum* C and D phage DNAs, which also carry the neurotoxin C1 and D genes (Popoff *et al.* 1990; 1991; Hauser *et al.* 1993) (EMBL Data Library access numbers for C3 genes: X59039 for *C. botulinum* C and X59040 for *C. botulinum* D).

Purification and sources

The classical method to obtain C3 is to cultivate anaerobically, in TGY medium, *C. botulinum* D strain 1873 or C strain ATCC 17784 and to follow for purification the procedure described by Rubin *et al.* (1988) or Aktories

et al. (1987). These methods have been made obsolete by the fact that recombinant C3 in *E. coli* can be obtained easily, in large quantities, and without C2 toxin contamination. Two main methods have been proposed to produce recombinant C3:

1. From the C3 gene cloned in an expression vector (Popoff *et al.* 1991), followed by a classical purification of C3 (Aktories *et al.* 1987; Rubin *et al.* 1988).

2. From a glutathione-S-transferase (GST) C3 fusion (Nobes and Hall 1995), and from a poly-Histidine tag C3 fusion (Lemichez *et al.* manuscript in preparation). For GST-C3 or His-C3 fusions, purification of recombinant C3 is obtained easily by affinity chromatography.

C3 purified from *C. botulinum* can be purchased from LIST Biological Laboratories, 501-B Vandell Way, Cambell, CA, 95008-6967, USA (fax: (408) 866-6364) or from BIOMOL Research Laboratories (ref catalog G-130), 5166 Campus Drive, Plymouth Meeting, PA, 19462, USA (fax: (610) 941 9252). Rabbit polyclonal antibodies against C3 are obtained easily upon immunization with a classical procedure. These antibodies can be used to immunoblot or neutralize C3.

Use in cell biology

C3 is used to selectively inactivate the small GTP-binding protein Rho. Although C3 can be used with [32P]-NAD to label *in vitro* or inactivate Rho proteins from cell lysates or in reconstituted systems (at a concentration of 1ng μl^{-1} in 20 mM HEPES buffer pH 8.00, containing 5 μM [32P]-NAD; presence of high salt concentrations and EDTA decrease the C3 enzymatic activity), C3 does not penetrate into cells. Several techniques have thus been proposed to inactivate Rho by C3 in cultured cells. The first one is to incubate cells with high concentrations of C3 (3–40 μg/ml depending on cell types). This method, however, remains expensive in term of C3 amounts. Microinjection of C3 is a classical method which is performed by injecting C3 into the cell, usually at a concentration of 100 μg ml^{-1} in the injecting pipette (Nobes and Hall 1995). A third method is to use a recombinant chimeric protein made by the fusion of the C3 gene with the B portion of diphtheria toxin (DT) DNA. This chimeric toxin named C3B can be used on cells sensitive to DT (mouse and rat cells are resistant to DT) at concentrations of about 10^{-8} M (Aullo *et al.* 1993). A recently published method is to transfect cells with a eukaryotic vector, expressing the C3 gene (N-terminally C-myc epitope tagged C3 transferase) (Hill *et al.* 1995). Introduction of C3 into cells is followed more or less rapidly (a few minutes by microinjection, hours by the chimeric C3B toxin) by morphological changes due to inactivation of the Rho GTP-binding proteins. The first sign of intoxication is a rounding of the cell body with formation of arborescent extensions. Staining of the F-actin cell structures with FITC-phalloidin, shows that C3 induces a total disruption of actin stress fibres indicating that Rho is a major player in the regulation of focal adhesion points (Chardin *et al.* 1988; Ridley and Hall 1992). Recently, Rho has been shown to regulate a phosphatidyl inositol 4-phosphate 5-kinase (Chong *et al.* 1994) producing 4,5-PIP2 (a phosphorylated inositol lipid required in the activation of proteins such as the actin-binding protein profilin or, involved in the intracellular traffic such as phospholipase D). Rho also plays a role in the signalling cascade triggered by growth factors such as lysophosphatidic acid (LPA) leading to transcriptional activation of SRE by SRF (Hill *et al.* 1995). In all these cases, C3 has been used to demonstrate the role of Rho in these reactions.

■ References

Aktories, K., Weller, U., and Chaatwal, G. S. (1987). *Clostridium botulinum* type C produces a novel ADP-ribosyl transferase distinct from botulinum C2 toxin. *FEBS Lett.*, **212**, 109–13.

Aktories, K., Braun, U., Rösener, S., Just, I. and Hall, A. (1989). The *rho* gene product expressed in *E. coli* is a substrate of botulinum ADP-ribosyl transferase C3. Biochem. *Biophys. Res.Comm.*, **158**, 209–13.

Aullo, P., Giry, M., Olsnes, S., Popoff, M. R., Kocks, C., and Boquet, P. (1993). A chimeric toxin to study the role of the p21 GTP-binding protein Rho in the control of actin microfilament assembly. *EMBO J.*, **12**, 921–31.

Belyi, Y. F., Tatakovskii, I. S., Vertief, Y. V., and Prosorovskii, S. V. (1991). Partial purification and characterization of an ADPribosyltransferase produced by *Legionella pneumophila*. *Biomed. Sci.*, **2**, 169–74.

Chardin, P., Boquet, P., Madaule, P., Popoff, M. R., Rubin, E. J., and Gill, D. M. (1989). The mammalian G-protein Rho C is ADP-ribosylated by *Clostridium botulinum* exoenzyme C3 and affects actin microfilaments in Vero cells. *EMBO J.*, **8**, 1087–92.

Chong, L. D., Traynor-Kaplan, A., Bokoch, G. M., and Schwartz, M. A. (1994). The small GTP-binding protein Rho regulates a phosphatidylinositol 4-phosphate 5-kinase in mammalian cells. *Cell*, **79**, 507–513.

Hall, A. (1994). GTPases and the actin cytoskeleton. *Ann. Rev. Cell Biol.*, **10**, 31–54.

Hauser, D., Gibert, M., Eklund, M. W., Boquet, P., and Popoff, M. R. (1993). Comparative analysis of C3 and botulinal neurotoxin genes and their environment in *Clostridium botulinum* types C and D. *J. Bacteriol.*, **175**, 7260–8.

Hill, C. S., Wynne, J. and Treisman, R. (1995). The Rho family GTPases RhoA, Rac1, and Cdc42Hs regulate transcriptional activation by SRF. *Cell*, **81**, 1159–70.

Inoue, S., Sugai, M., Muruooka, Y., Paik, S. Y., Hong, Y. M. Ohgai, H., *et al.* (1991). Molecular cloning and sequencing of the epidermal cell differentiation inhibitor gene from *Staphylocccus aureus*. *Biochem. Biophys. Res. Comm.*, **174**, 459–64.

Just, I., Mohrt, C., Schallehn, G., Menard, L., Didsbury, J. R., Vandekerckove, J., *et al.* (1992a). Purification and characterization of an ADP-ribosyltransferase produced by *Clostridium limosum*. *J. Biol. Chem.*, **267**, 10274–80.

Just, I., Schallenhn, G. and Aktories, K. (1992b). ADP-ribosylation of small GTP-binding proteins by *Bacillus cereus*. *Biochem. Biophys. Res. Comm.*, **183**, 931–6.

Narumiya, S. and Morii, N. (1993). *rho* gene products, botulinum C3 exoenzyme and cell adhesion. *Cellular Signalling*, **5**, 9–19.

Nobes, C. D. and Hall, A. (1995). Rho, Rac, and Cdc42 GTPases regulate the assembly of multimolecular complexes associated with actin stress fibers, lamellipodia, and filopodia. *Cell*, **81**, 53–62.

Popoff, M. R., Boquet, P., Gill, D. M., and Eklund, M. W. (1990). DNA sequence of exoenzyme C3, an ADP-ribosyltransferase encoded by *Clostridium botulinum* C and D phages. *Nucleic Acid Res.*, **18**, 1291.

Popoff, M. R., Hauser, D., Boquet, P., Eklund, M. W., and Gill, D. M. (1991). Characterization of the C3 gene of *Clostridium botulinum* types C and D and its expression in *E. coli*. *Infect. Immun.*, **59**, 3673–9.

Ridley, A. J. and Hall, A. (1992). The small GTP-binding protein Rho regulates the assembly of focal adhesions and actin stress fibers in response to growth factors. *Cell*, **70**, 389–99.

Rubin, E. J., Gill, D. M., Boquet, P., and Popoff, M. R. (1988). Functional modification of a 21 kiloDalton G-protein when ADP-ribosylated by exoenzyme C3 from *Clostridium botulinum*. *Mol. Cell Biol.*, **8**, 418–26.

Sekine, A., Fujiwara, M., and Narumiya, S. (1989). Asparagine residue in the *rho* gene product is the modification site for botulinum ADP-ribosyl transferase. *J. Biol. Chem.*, **264**, 8602–5.

■ *Emmanuel Lemichez, Patrice Boquet, and Michel R. Popoff:*
Unité des Toxines Microbiennes Institut Pasteur, 75724 Paris Cedex 15, France and
INSERM U452 Faculté de Médecine de Nice, 06107 Nice Cedex 2, France

Adenylate cyclase toxin (*Bordetella* sp.)

Adenylate cyclase toxin is a 177.7 kDa RTX protein secreted by pathogenic strains of Bordetella. *Upon post-translational fatty-acylation of an internal lysine, the toxin gains the capacity to insert into the plasma membrane of a wide variety of eukaryotic cells, to deliver its amino-terminal adenylate cyclase (AC) domain into their cytosol and to form small cation selective membrane channels. The AC domain binds intracellular calmodulin and converts ATP to cAMP, causing steep increase of intracellular cAMP. This results in disruption of cellular signalling and microbicidal capacities of immune effector cells.*

Adenylate cyclase toxin (ACT or CyaA) is a key virulence factor produced by the human whooping cough agents, *Bordetella pertussis* and *parapertussis* and by the related animal pathogens *B. bronchiseptica* and *B. avium* (Weiss and Hewlett 1986; Hanski and Coote 1991; Mock and Ullmann 1993)**.** The toxin is essential in the early stages of bacterial colonization of the respiratory tract (Goodwin and Weiss 1990) and can induce apoptosis of lung alveolar macrophages (Khelef *et al.* 1993). ACT is produced as a single 177.7 kDa polypeptide chain protoxin of 1706 residues (Glaser *et al.* 1988) (EMBL accession number: Y00545), which is activated by a post-translational palmitoylation of the internal lysine residue 983 (Hackett *et al.* 1994), requiring an accessory protein, CyaC (Barry *et al.* 1991). As shown in Fig. 1, the toxin is a bifunctional protein consisting of a cell-invasive and calmodulin-dependent adenylate cyclase (AC) catalytic domain (400 amino-terminal residues), fused to a pore-forming hemolysin consisting of 1306 residues. The AC enzyme penetrates into target cells, where it binds calmodulin and catalyses high-level synthesis of the key signal molecule, cAMP, thereby disrupting cellular functions (Confer and Eaton 1982). The hemolysin portion of ACT accounts for membrane insertion of ACT and AC-delivery into cells and can itself form small cation selective membrane channels permeable for calcium (Benz *et al.* 1994; Iwaki *et al.* 1995). These cause colloido-osmotic cell-lysis and account for the weak hemolytic activity of ACT that is independent of the presence and activity of the AC domain

(Sakamoto *et al.* 1992) The hemolysin is a typical RTX (**R**epeat in **ToX**in) protein consisting of a hydrophobic pore-forming domain, fatty-acylation domain, 38 calcium-binding glycine and aspartate-rich nonapeptide repeat domain, and a carboxy-terminal secretion signal (Welch 1991; Sebo and Ladant 1993).

The toxin contains about 40 calcium binding sites of various affinities located in the repeats and its activity is calcium dependent (Hanski and Farfel 1985; Hewlett *et al.* 1991; Rose *et al.* 1995). The target cell(s) for ACT *in vivo* was not defined and is popularly believed to be alveolar macrophages and leukocytes. There is indirect evidence that delivery of AC domain into cells can be accomplished by toxin monomers (Fig. 2), while oligomerization of ACT is involved in formation of membrane channels (Sakamoto *et al.* 1992; Betsou *et al.* 1993; Szabo *et al.* 1994).

■ Purification and sources

ACT is extracted from the outer membrane of exponentially growing Bordetella organisms with buffered 4 M urea. Alternatively, recombinant ACT is extracted with buffered 8 M urea from cell debris of *Escherichia coli* strains expressing *cyaA* and *cyaC* genes (Sebo *et al.* 1991; Betsou *et al.* 1993). Nearly homogeneous toxin preparations are obtained by various combinations of sucrose density gradient centrifugation and chromatographies on

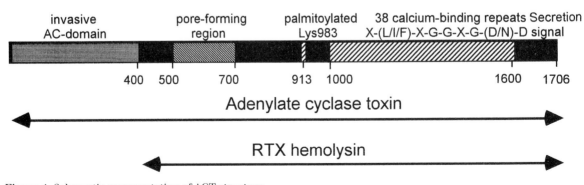

Figure 1. Schematic representation of ACT structure.

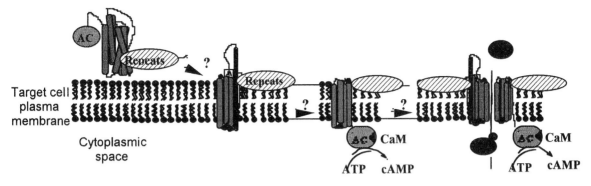

Figure 2. Schematic representation of ACT action.

hydrophobic, ion-exchange, and Calmodulin–Agarose affinity resins (Hewlett *et al.* 1989; Rogel *et al.* 1989; Gentile *et al.* 1990; Sakamoto *et al.* 1992). Purified toxin is diluted and stored frozen in 8 M urea and directly applied into urea free assay buffers containing target cells and millimolar calcium. Upon dialysis, the purified toxin rapidly loses activity. Intact toxin migrates abberantly on SDS-PAGE as a 220 kDa band. ACT cannot yet be purchased from commercial suppliers and submilligram amounts of toxin purified from *B. pertussis* can be provided by Drs E. Hanski (Hebrew University–Hadassah Medical School, Jerusalem), E. L. Hewlett (University of Virginia School of Medicine, Charlottesville), and J. G. Coote (Glasgow University, Scotland). Larger amounts of recombinant toxin can be obtained from Drs Agnes Ullmann (Institut Pasteur, Paris) and P. Sebo (Institute of Microbiology, Czech Acad. Sci., Prague).

Toxicity

Toxicity has not been evaluated. Mice tolerate intraperitoneal injections of 2.5 mg/kg without observable difficulties (C. Leclerc, personal communication). No problems to operators handling hundreds of milligrams of active toxin were observed. In addition, vaccination with active ACT protects mice against colonization by *B. pertussis* (Betsou *et al.* 1993).

Use in cell biology

Translocation of ACT into cells is strictly calcium dependent, with half maximal activity at 0.6 mM calcium concentration (Hanski and Farfel 1985; Hewlett *et al.* 1991). The toxin has the capacity to insert into the plasma membrane of a broad spectrum of eukaryotic cell types, including immune effector cells, and to deliver its AC domain into the cell interior by an unknown mechanism, directly across the plasma membrane and without the need for endocytosis (Gordon *et al.* 1989). ACT action does not appear to involve interaction with any specific proteinaceous receptor on the cell surface, and membrane insertion of ACT is apparently preceded by a rather

unspecific adsorption to cell surface (Iwaki *et al.* 1995). Translocation of AC across membrane is stimulated by negative transmembrane potential (Otero *et al.* 1995). Inside nucleated cells, the AC enzyme is rapidly inactivated by an ATP-dependent mechanism (Gilboa-Ron *et al.* 1989). Because of its cell-invasive activity, recombinant ACT with inserted heterologous epitopes was used for delivery of viral epitopes into MHC class I, restricted antigen presentation pathway and induction of specific CD8[+] cytotoxic T lymphocytes in mice (Fayolle *et al.* 1996; Sebo *et al.* 1995).

The capacity of ACT to cause steep and transient increase of intracellular cAMP was used for reversibly maintaining meiotic arrest of rat oocytes (Aberdam *et al.* 1987).

References

Aberdam, E., Hanski, E., and Dekel, N. (1987). Maintenance of meiotic arrest in isolated rat oocytes by the invasive adenylate cyclase of *Bordetella pertussis*. *Biol. Reprod.*, **36**, 530–5.

Barry, E. M., Weiss, A. A., Ehrmann, I. E., Gray, M. C., Hewlett, E. L., and Goodwin, M. S. (1991). *Bordetella pertussis* adenylate cyclase toxin and hemolytic activities require a second gene, *cyaC*, for activation. *J. Bacteriol.*, **173**, 720–6.

Benz, R., Maier, E., Ladant, D., Ullmann, A., and Sebo, P. (1994). Adenylate cyclase toxin of *Bordetella pertussis* evidence for the formation of small ion-permeable channels and comparison with HlyA of *Escherichia coli*. *J. Biol. Chem.*, **269**, 27231–9.

Betsou, F., Sebo, P., and Guiso, N. (1993). CyaC-mediated activation is important not only for toxic but also for protective activities of *Bordetella pertussis* adenylate cyclase-hemolysin. *Infect. Immun.*, **61**, 3583–9.

Confer, D. L. and Eaton, J. W. (1982). Phagocyte impotence caused by an invasive bacterial adenylate cyclase. *Science*, **217**, 948–50.

Fayolle, C., Sebo, P., Ladant, D., Ullmann, A., and Leclerc, C. (1996). *In vivo* induction of CTL responses by recombinant adenylate cyclase of *Bordetella pertussis* carrying viral CD8+ T-cell epitope. *J. Immunol.*, **156**, 4697–706.

Gentile, F., Knipling, L. G., Sackett, D. L., and Wolff, J. (1990). Invasive adenylyl cyclase of *Bordetella pertussis*. *J. Biol. Chem.*, **265**, 10686–92.

Gilboa-Ron, A., Rogel, A., and Hanski, E. (1989). *Bordetella pertussis* adenylate cyclase inactivation by the host cell. *Biochem. J.*, **262**, 25–31.

Glaser, P., Ladant, D., Sezer, O., Pichot, F., Ullmann, A., and Danchin, A. (1988). The calmodulin-sensitive adenylate cyclase of *Bordetella pertussis*: cloning and expression in *Escherichia coli*. *Mol. Microbiol.*, **2**, 19–30.

Goodwin, M. S. and Weiss, A. A. (1990). Adenylate cyclase toxin is critical for colonization and pertussis toxin is critical for lethal infection by *Bordetella pertussis* in infant mice. *Infect. Immun.*, **58**, 3445–7.

Gordon, V. M., Young Jr, W. W., Lechler, S. M., Gray, M. C., Leppla, S. H., and Hewlett, E. L. (1989). Adenylate cyclase toxins from *Bacillus anthracis* and *Bordetella pertussis*. Different processes for interaction with and entry into target cells. *J. Biol. Chem.*, **264**, 14792–6.

Hackett, M., Guo, L., Shabanowitz, J., Hunt, D. F., and Hewlett, E. L. (1994). Internal lysine palmitoylation in adenylate cyclase toxin from *Bordetella pertussis*. *Science*, **266**, 433–5.

Hanski, E. and Coote, J. G. (1991) *Bordetella pertussis* adenylate cyclase toxin. In *Sourcebook of bacterial toxins* (ed. J. E. Alouf and J. H. Freer), pp. 349–366, Academic Press, New York.

Hanski, E. and Farfel, Z. (1985). *Bordetella pertussis* invasive adenylate cyclase: partial resolution and properties of its cellular penetration. *J. Biol. Chem.*, **260**, 5526–32.

Hewlett, E. L., Gordon, V. M., McCaffery, J. D., Sutherland, W. M., and Gray, M. C. (1989). Adenylate cyclase toxin from *Bordetella pertussis*. Identification and purification of the holotoxin molecule. *J. Biol. Chem.*, **264**, 19379–84.

Hewlett, E. L., Gray, L., Allietta, M., Ehrmann, I. E., Gordon, V. M., and Gray, M. C. (1991). Adenylate cyclase toxin from *Bordetella pertussis*. Conformational change associated with toxin activity. *J. Biol. Chem.*, **266**, 17503–8.

Iwaki, M., Ullmann, A., and Sebo, P. (1995). Identification by *in vitro* complementation of regions required for cell-invasive activity of *Bordetella pertussis* adenylate cyclase toxin. *Mol. Microbiol.*, **17**, 1015–24.

Khelef, N., Zychlinsky, A., and Guiso, N. (1993). *Bordetella pertussis* induces apoptosis in macrophages: role of adenylate cyclase-hemolysin. *Infect. Immun.*, **61**, 4064–70.

Mock, M. and Ullmann, A. (1993) Calmodulin-activated bacterial adenylate cyclases as virulence factors. *Trends Microbiol.*, **1**, 187–92.

Otero, A. S., Yi, X., Gray, M. C., Szabo, G., and Hewlett, E. L. (1995). Membrane depolarization prevents cell invasion by *Bordetella pertussis* adenylate cyclase toxin. *J. Biol. Chem.*, **270**, 9695–7.

Rogel, A., Schultz, J., Brownlie, R. M., Coote, J. G., Parton, R., and Hanski, E. (1989). *Bordetella pertussis* adenylate cyclase: purification and characterization of the toxic form of the enzyme. *EMBO J.*, **8**, 2755–60.

Rose, T., Sebo, P., Bellalou, J., and Ladant, D. (1995). Interaction of calcium with *Bordetella pertussis* adenylate cyclase toxin: Characterization of multiple calcium-binding sites and calcium induced conformational changes. *J. Biol. Chem.*, **270**, 26370–6.

Sakamoto, H., Bellalou, J., Sebo, P. and Ladant, D. (1992). *Bordetella pertussis* adenylate cyclase toxin: structural and functional independence of the catalytic and hemolytic activities. *J. Biol. Chem.*, **267**, 13598–602.

Sebo, P. and Ladant, D. (1993). Repeat sequences in the *Bordetella pertussis* adenylate cyclase toxin can be recognized as alternative carboxy-proximal secretion signals by the *Escherichia coli* a-haemolysin translocator. *Mol. Microbiol.*, **9**, 999–1009.

Sebo, P., Glaser, P., Sakamoto, H., and Ullmann, A. (1991). High-level synthesis of active adenylate cyclase toxin of *Bordetella pertussis* in a reconstructed *Escherichia coli* system. *Gene*, **104**, 19–24.

Sebo, P., Fayolle, P., d'Andria, O., Ladant, D., Leclerc, C., and Ullmann, A. (1995). Cell-invasive activity of the epitope-tagged adenylate cyclase of *Bordetella pertussis* allows *in vitro* delivery of foreign epitope into MHC class-I-restricted antigen processing pathway. *Infect Immun.*, **63**, 3851–7.

Szabo, G., Gray, M. C., and Hewlett, E. L. (1994). Adenylate cyclase toxin from *Bordetella pertussis* produces ion conductance across artificial lipid bilayers in a calcium and polarity-dependent manner. *J. Biol. Chem.*, **269**, 22496–9.

Weiss, A. A. and Hewlett, E. L. (1986) Virulence factors of *Bordetella pertussis*. *Ann. Rev. Microbiol.*, **40**, 661–86.

Welch, R. A. (1991). Pore-forming cytolysins of gram-negative bacteria. *Mol. Microbiol.*, **5**, 521–8.

■ *Peter Sebo* and Agnes Ullmann:*
Unité de Biochimie des Régulations Cellulaires,
Institut Pasteur,
28 rue du Docteur Roux,
75015 Paris Cedex 15,
France
*Present address: Cell and Molecular Microbiology Division Institute of Microbiology of the Czech Academy of Sciences, Videnska 1083, CZ-142 Praha 4

Anthrax edema factor (*Bacillus anthracis*)

Anthrax toxin edema factor is an 'invasive adenylyl cyclase' which is internalized into nearly all types of cells by interaction with the protective antigen toxin component. Translocation to the cytosol leads to unregulated conversion of ATP to cAMP, which activates cAMP-dependent protein kinase, leading to serious perturbations of regulatory mechanisms.

Anthrax edema factor (EF, originally designated Factor I) is a calmodulin-dependent adenylyl cyclase (Leppla 1982) which is one of the three proteins of the anthrax toxin complex (Smith *et al.* 1955; Leppla 1991*a*; 1995). The EF gene (Escuyer *et al.* 1988; Mock *et al.* 1988; Robertson *et al.* 1988; Tippetts and Robertson 1988) (GenBank accessions M23179 and M24074), located on the large pXO1 plasmid, encodes a precursor of 800 amino acids. Cleavage of the 33-amino acid signal peptide produces the 767 residue mature protein (PIR Protein database accession JS0029) having a mass of 88.8 kDa. The other

proteins of the toxin complex are protective antigen (PA, GenBank M22589) and lethal factor (LF, GenBank M29081 and M30210; see separate entry in this book p. 00). None of the three proteins contain cysteine. As shown in Fig. 1, EF enters cells by binding to proteolytically activated, receptor-bound PA (Klimpel *et al.* 1992). EF (or LF) is then endocytosed and translocated from endosomes directly to the cytosol of cells, where it produces unregulated high concentrations of cAMP (Leppla 1982; Gordon *et al.* 1988). Entry of EF and LF is totally dependent on PA. The receptor for PA is unknown, but is present on nearly all types of cells (Escuyer and Collier 1991; Leppla 1991*a*).

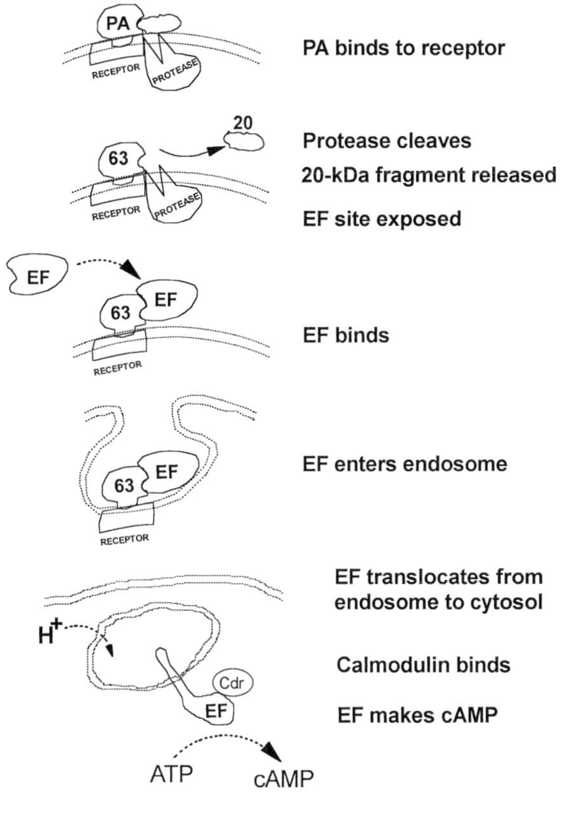

EF has two recognized domains, an N-terminal region that causes binding to PA and internalization, and the C-terminal catalytic domain (Escuyer *et al.* 1988; Robertson 1988). Residues 1–250 constitute the PA-binding domain; this has strong sequence homology to LF. Fusion of residues 1–254 of LF or 1–260 of EF to the catalytic domains of other toxins (e.g. diphtheria, tetanus) produces potent cytotoxins, due to PA-dependent delivery to the cytosol (Arora and Leppla 1993). Residues 265–570 of EF have sequence homology to the other known 'invasive adenylate cyclase', that of *Bordetella pertussis* (Escuyer *et al.* 1988; Robertson 1988). Expression of residues 262–767 of EF in *Escherichia coli* yields a fully active adenylyl cyclase (Labruyere *et al.* 1991).

Several regions of EF have been shown to be involved in catalysis. Residues 314–321, GLNEHGKS, fit the consensus GxxxxGKS sequence of ATP binding sites; substitution of K320 destroys enzyme activity (Xia and Storm 1990). Identification of the EF sequences comprising the site to which calmodulin binds has been difficult. A synthetic peptide corresponding to EF residues 499–532 binds tightly to calmodulin (Munier *et al.* 1993), suggesting this region is part of the calmodulin-binding site.

Figure 1. Schematic representation of the edema factor intoxication of eukaryotic cells.

PA binds to receptor

Protease cleaves
20-kDa fragment released
EF site exposed

EF binds

EF enters endosome

EF translocates from endosome to cytosol

Calmodulin binds

EF makes cAMP

■ Purification and sources

EF (and also PA and LF) are secreted by strains of *B. anthracis* containing plasmid pXO1. Mutant strains have been made which produce only EF, PA, or LF (Pezard *et al.* 1993). The three proteins collectively constitute more than half of the protein in a culture supernatant. Purification has been accomplished by ammonium sulfate precipitation, followed by chromatography on hydroxy-apatite, DEAE, and other anion exchange resins (Quinn *et al.* 1988; Leppla 1991*b*). Hydroxyapatite is an effective step because EF eluted last, well separated from the other components. EF is susceptible to proteolysis and is the most difficult of the three components to produce. Fermentor cultures yield 10 mg PA, 2 mg LF, and 0.5 mg EF per litre (Leppla 1991*b*). Purified PA should be examined by electrophoresis to screen for cleavage at residue 313, which destroys activity (Leppla 1991*a*). The proteins are not available commercially. Certain toxin components may be available from Stephen Leppla (NIDR, NIH, Bethesda, MD), Michelle Mock (Pasteur Institute, Paris), Peter Turnbull (CAMR, Porton Down, UK), or Joseph Farchaus (USAMRIID, Frederick, MD; PA only).

■ Toxicity

The edema toxin (PA + EF) was originally detected using vascular permeability assays in guinea pig skin (Smith *et al.* 1955). More convenient is measurement of elevated cytosolic cAMP concentrations in cultured cells (Leppla 1982; Gordon *et al.* 1988). Nearly all types of cells respond, although intracellular cAMP concentrations produced by edema toxin differ widely among cultured cells, probably due to differences in PA receptor number, cytosolic stability of EF, concentrations of Ca^{2+}-activated calmodulin, or activity of cAMP phosphodiesterase. Some cell types demonstrate 1000-fold increases, reaching 2000 μmol per mg cell protein (Leppla 1982; Gordon *et al.* 1989), which may represent conversion of 20–50 per cent of the cellular ATP. Elevated cAMP impairs phagocytic cells and thereby contributes to establishment of *B. anthracis* infections. Because all the effects of cAMP appear to be mediated by cAMP-dependent protein kinase, and this is fully activated by small increases in intracellular cAMP, no additional sequelae are expected from generation of very high concentrations of cAMP, except those caused by depletion of ATP. Elevated cAMP inhibits cell growth but is not lethal to most types of cells, and cultured cells generally will resume growth after toxin is removed. The anthrax toxin components are individually nontoxic and therefore present little hazard to users.

■ Catalytic properties

EF is a very efficient adenylyl cyclase (Labruyere *et al.* 1991; Leppla 1991*a*), having K_m for Mg^{+2}-ATP = 0.16 mM, and V_{max} = 1.2 mmol cAMP/min/mg enzyme. The catalytic activity is absolutely dependent on calmodulin and highly dependent on calcium. The truncated recombinant protein containing only residues 262–767 (Labruyere *et al.* 1991) has been useful in studying catalytic properties. This 62-kDa product has kinetic activities like those of the full-size protein, with a K_m = 0.25 mM for ATP, and an absolute requirement for calmodulin. The K_d for calmodulin activation is 23 nM.

■ Use in cell biology

Edema toxin (PA + EF) provides a convenient reagent for elevating cAMP concentrations in cultured cells. The effect of the toxin is qualitatively the same as that produced by cholera toxin, but the response occurs in a wider variety of cell types and higher concentrations of cAMP are typically produced. A further advantage as compared to cholera toxin is that cAMP concentrations decrease rapidly upon toxin removal, apparently because the toxin is inactivated within cells. It was estimated that the half-life of EF in CHO cells is less than 2 hours (Leppla 1982).

The high catalytic activity of EF suggests it could be used to synthesize cAMP analogues. If the enzyme can be shown to accept ATP analogues as substrates, then certain cAMP analogues could be prepared that might have useful pharmacological properties.

■ References

Arora, N. and Leppla, S. H. (1993). Residues 1–254 of anthrax toxin lethal factor are sufficient to cause cellular uptake of fused polypeptides. *J. Biol. Chem.*, **268**, 3334–41.

Escuyer, V. and Collier, R. J. (1991). Anthrax protective antigen interacts with a specific receptor on the surface of CHO-K1 cells. *Infect. Immun.*, **59**, 3381–6.

Escuyer, V., Duflot, E., Sezer, O., Danchin, A., and Mock, M. (1988). Structural homology between virulence-associated bacterial adenylate cyclases. *Gene*, **71**, 293–8.

Gordon, V. M., Leppla, S. H., and Hewlett, E. L. (1988). Inhibitors of receptor-mediated endocytosis block the entry of *Bacillus anthracis* adenylate cyclase toxin but not that of *Bordetella pertussis* adenylate cyclase toxin. *Infect. Immun.*, **56**, 1066–9.

Gordon, V. M., Young, W. W., Jr, Lechler, S. M., Gray, M. C., Leppla, S. H., and Hewlett, E. L. (1989). Adenylate cyclase toxins from *Bacillus anthracis* and *Bordetella pertussis*. Different processes for interaction with and entry into target cells. *J. Biol. Chem.*, **264**, 14792–6.

Klimpel, K. R., Molloy, S. S., Thomas, G., and Leppla, S. H. (1992). Anthrax toxin protective antigen is activated by a cell-surface protease with the sequence specificity and catalytic properties of furin. *Proc. Natl. Acad. Sci. USA*, **89**, 10277–81.

Labruyere, E., Mock, M., Surewicz, W. K., Mantsch, H. H., Rose, T., Munier, H., *et al.* (1991). Structural and ligand-binding properties of a truncated form of *Bacillus anthracis* adenylate cyclase and of a catalytically inactive variant in which glutamine substitutes for lysine-346. *Biochemistry*, **30**, 2619–24.

Leppla, S. H. (1982). Anthrax toxin edema factor: a bacterial adenylate cyclase that increases cyclic AMP concentrations of eukaryotic cells. *Proc. Natl. Acad. Sci. USA*, **79**, 3162–6.

Leppla, S. H. (1991*a*). The anthrax toxin complex. In *Sourcebook of bacterial protein toxins* (ed. J. E. Alouf and J. H. Freer), pp. 277–302, Academic Press, London.

Leppla, S. H. (1991*b*). Purification and characterization of adenylyl cyclase from *Bacillus anthracis*. *Methods Enzymol.*, **195**, 153–68.

Leppla, S. (1995). Anthrax toxins. In *Bacterial toxins and virulence factors in disease. Handbook of natural toxins* Vol. 8 (ed. J. Moss, B. Iglewski, M. Vaughan, and A. Tu), pp. 543–72, Marcel Dekker, New York.

Mock, M., Labruyere, E., Glaser, P., Danchin, A., and Ullmann, A. (1988). Cloning and expression of the calmodulin-sensitive

Bacillus anthracis adenylate cyclase in *Escherichia coli*. *Gene*, **64**, 277–84.

Munier, H., Blanco, F. J., Precheur, B., Diesis, E., Nieto, J. L., Craescu, C. T., *et al.* (1993). Characterization of a synthetic calmodulin-binding peptide derived from *Bacillus anthracis* adenylate cyclase. *J. Biol. Chem.*, **268**, 1695–701.

Pezard, C., Duflot, E., and Mock, M. (1993). Construction of *Bacillus anthracis* mutant strains producing a single toxin component. *J. Gen. Microbiol.*, **139**, 2459–63.

Quinn, C. P., Shone, C. C., Turnbull, P. C., and Melling, J. (1988). Purification of anthrax-toxin components by high-performance anion-exchange, gel-filtration and hydrophobic-interaction chromatography. *Biochem. J.*, **252**, 753–8.

Robertson, D. L. (1988). Relationships between the calmodulin-dependent adenylate cyclases produced by *Bacillus anthracis* and *Bordetella pertussis*. *Biochem. Biophys. Res. Commun.*, **157**, 1027–32.

Robertson, D. L., Tippetts, M. T., and Leppla, S. H. (1988). Nucleotide sequence of the *Bacillus anthracis* edema factor gene (cya): a calmodulin-dependent adenylate cyclase. *Gene*, **73**, 363–71.

Smith, H., Keppie, J., and Stanley, J. L. (1955). The chemical basis of the virulence of *Bacillus anthracis*. V. the specific toxin produced by *B. anthracis in vivo*. *Brit. J. Exp. Path.*, **36**, 460–72.

Tippetts, M. T. and Robertson, D. L. (1988). Molecular cloning and expression of the *Bacillus anthracis* edema factor toxin gene: a calmodulin-dependent adenylate cyclase. *J. Bacteriol.*, **170**, 2263–6.

Xia, Z. G. and Storm, D. R. (1990). A-type ATP binding consensus sequences are critical for the catalytic activity of the calmodulin-sensitive adenylyl cyclase from *Bacillus anthracis*. *J. Biol. Chem.*, **265**, 6517–20.

■ *Stephen H. Leppla:*
Laboratory of Microbial Ecology,
National Institute of Dental Research,
NIH, Bethesda, MD 20892, USA

3

Toxins affecting protein synthesis

Introduction

During the past decade there has been an enormous increase in our understanding of the structure–function relationships and molecular mechanism of action of many of the toxins that affect protein synthesis. The toxins considered in this section inhibit eukaryotic cellular protein synthesis either by the NAD^+ dependent adenosine diphosphate ribosylation of elongation factor 2 or function as an N-glycosidase to remove adenine 4324 (A^{4324}) from 28S rRNA which leads to an impaired ability of the ribosome to bind elongation factors. The X-ray crystal structure of ricin (Montfort et al. 1987; Rutenber et al. 1991), *Pseudomonas* exotoxin A (Allured et al. 1986), diphtheria toxin (Choe et al. 1992; Bennett et al. 1994), and Shiga toxin (Fraser et al. 1994) have been solved. Both crystallographic and biochemical genetic analysis have led to a detailed understanding of the adenosine diphosphate ribosyltransferase activity of diphtheria toxin and *Pseudomonas* exotoxin A (Collier 1990; Wick and Iglewski 1990) and the N-glycosidase activity of ricin, Shiga toxin, and the plant derived ribosome-inactivating proteins (Jimenez and Vasquez 1985; Stirpe and Barbieri 1986). Most all of these toxins have been highly purified, and their respective genes have been cloned and sequenced. In addition, many of their respective enzymatically active fragments or chains have been employed in the assembly of immunotoxins by the chemical crosslinking of the catalytic domains of either a bacterial or plant toxin to a monoclonal antibody (Ghetie and Vitetta 1994). More recently, recombinant DNA and protein engineering methods have been used to create chimeric genes in which the native receptor binding domain of either diphtheria toxin or *Pseudomonas* exotoxin A have been genetically substituted with one of a variety of genes encoding a polypeptide ligand that is directed toward a specific cell surface receptor or antigenic determinant on the target cell surface (for reviews see Kreitman and Pastan 1994; Murphy et al. 1995). The immunotoxins and the protein fusion toxins represent novel reagents for the study of cell biology and for the experimental treatment of human disease. Indeed, the diphtheria toxin-related IL-2 fusion toxin DAB_{389} IL-2 is currently being evaluated in Phase III clinical trials for the treatment of cutaneous T cell lymphoma.

References

Allured, V. S., Collier, R. J., Carroll, S. F., and McKay, D. B. (1986). Structure of exotoxin A of *Pseudomonas aeruginosa* at 3.0-Angstorm resolution. *Proc. Natl. Acad. Sci. USA*, **83**, 1320–4.

Bennett, M. J., Choe, S., and Eisenberg, D. (1994). Domain swapping entangling alliances between proteins. *Proc. Natl. Acad. Sci. USA*, **91**, 3127–31.

Choe, S., Bennett, M. J., Fujii, G., Curmi, P. M. G., Kantardjieff, K. A., Collier, R. J. et al. (1992). The crystal structure of diphtheria toxin. *Nature*, **357**, 216–22.

Collier, R. J. (1990). In ADP-ribosylating toxins and g proteins (ed. J. Moss and M. Vaughan) pp. 3–19, American Society for Microbiology, Washington, DC.

Fraser, M. E., Chernaia, M. M., Kozlov, Y. V., and James, M. N. G. (1994). Crystal structure of the holotoxin from Shigella dysenteriae at 2. 5. resolution. *Nature Structural Biol.*, **1**, 59–64.

Ghetie, V. and Vitetta, E. (1994). Immunotoxins in the therapy of cancer: from bench to clinic. *Pharmacology & Therapeutics*, **63**, 209–34.

Jimenez, A. and Vazquez, D. (1985). Plant and fungal protein glycoprotein toxins inhibiting eukaryote protein synthesis. (*Annu.*) *Rev. Microbiol.*, **39**, 649–72.

Kreitman, R. J. and Pastan, I. (1994). Recombinant single-chain immunotoxins against T and B cell leukemias. *Leukemia & Lymphoma*, **13**, 1–10.

Montfort, W., Villafranca, J. E., Monzingo, A. F., Ernst, S., Katzin, B., Rutenber, E., et al. (1987). The three-dimensional structure of ricin at 2. 8 A. *J. Biol. Chem.*, **262**, 5389–403.

Murphy, J. R., vanderSpek, J. C., Lemichez, E., and Boquet P. (1995). In *Bacterial toxins, virulence factors and disease*, vol. 8 (ed. J. Moss, M. Vaughan, B. Iglweski, and A, Tu), pp. 23–45, Marcel Dekker, New York.

Rutenber, E., Katzin, B. J., Collins, E. J., Mlsna, D., Ernst, S. E., Ready, M. P., et al. (1991). Crytallographic refinement of ricin to 2. 5 A. *Proteins*, **10**, 240–50.

Stirpe, F. and Barbieri, L. (1986). Ribosome-inactivating proteins up to date. *FEBS Lett.*, **195**, 1–8.

Wick, M. J., and Iglewski, B. H. (1990). In *ADP-ribosylating toxins and g proteins* (ed. J. Moss and M. Vaughan), pp. 31–43, American Society for Microbiology, Washington,DC.

■ *John R. Murphy:*
Evans Department of Clinical Research and Department of Medicine,
Boston University Medical Center Hospital,
Boston, MA 02118-2393,
USA

Diphtheria toxin (*Corynebacterium diphtheriae*)

Diphtheria toxin is a 535 amino acid protein that is produced and secreted into the growth medium by toxigenic strains of Corynebacterium diphtheriae. *The toxin is a three domain protein whose X-ray crystal structure has been recently described (Choe et al. 1992) and refined (Bennett et al. 1994). Diphtheria toxin binds to its eukaryotic cell surface receptor, a heparin binding epidermal growth factor-like precursor, and is internalized by receptor mediated endocytosis. Intact toxin is cleaved by the endoprotease furin in the 14 amino acid protease sensitive loop between the catalytic and transmembrane domains and following acidification of the endosome, the transmembrane domain of the toxin inserts into the vesicle membrane and facilitates the delivery of the catalytic domain to the cytosol. Once delivered to the cytosol the catalytic domain specifically ADP-ribosylates cellular elongation factor 2 which results in the inhibition of protein synthesis and death of the cell.*

Diphtheria toxin is the primary virulence factor of toxigenic *C. diphtheriae* the etiologic agent of clinical diphtheria (Pappenheimer 1977). The structural gene for diphtheria toxin, *tox*, is carried by a closely related family of corynebacteriophages of which the β-phage has been the best studied (Buck *et al.* 1985; Bishai and Murphy 1988). The regulation of *tox* gene expression is controlled by the *C. diphtheriae* determined iron-activated repressor DtxR (Tao *et al.* 1994). Diphtheria toxin is produced in maximal yield only during the decline phase of the bacterial growth cycle when iron becomes the growth rate limiting substrate. The toxin is synthesized in precursor form and is co-translationally secreted into the growth medium (Smith *et al.* 1980). As shown in Fig. 1, diphtheria toxin is a three domain protein and is composed of the catalytic (C), transmembrane (T), and receptor binding (R) domains. The R domain has been shown to specially bind to a heparin binding epidermal growth factor-like precursor on the surface of sensitive eukaryotic cells (Naglich *et al.* 1992).

Diphtheria toxin is readily cleaved into two polypeptide fragments following mild digestion with trypsin or other serine proteases (Drazin *et al.* 1971; Gill and Dinius 1971). After cleavage, the toxin may be separated under denaturing conditions in the presence of a thiol into two polypeptides. The N-terminal polypeptide (catalytic domain or fragment A; M_r 21 167) is an enzyme that catalyses the NAD^+-dependent ADP-ribosylation of elongation factor 2 (EF2) according to the following reaction:

fragment A (catalytic domain)
$$NAD^+ + EF2 \rightarrow ADPR\text{-}EF2 + nicotinamide + H^+$$

Once EF2 is ADP-ribosylated it can no longer participate in protein synthesis. The C-terminal polypeptide (transmembrane and receptor binding domains or fragment B; M_r 37 199) is required for binding the toxin to its cell surface receptor on sensitive cells and for facilitating the transport of fragment A into the cytosol. The nucleic acid

sequences of the native diphtheria toxin structural genes from corynebacteriophage β and ω have been shown to be identical (Greenfield *et al.* 1983; Ratti *et al.* 1983).

■ Purification and sources

Diphtheria toxin may be partially purified from the culture supernatant fluids of toxigenic *C. diphtheriae* by ammonium sulfate precipitation (65 per cent saturation) and ion exchange chromatography on DE-52 (Whatman) (Pappenheimer *et al.* 1972). In addition, diphtheria toxin in either its 'intact' or 'nicked' (i.e. cleaved after Arg190, Arg192, and/or Arg193) forms may be obtained from a variety of commercial sources (e.g. [intact or un-nicked toxin] List Laboratories, Campbell, CA; [nicked toxin] CalBiochem, La Jolla, CA; Connaught Laboratories, Toronto, Canada). In general, these preparations of toxin contain relatively high levels of contaminating nuclease activity. Diphtheria toxin may be further purified by high performance liquid chromatography (HPLC) gel filtration using a TSK-3000 column (Bodley *et al.* 1990).

■ Toxicity

Diphtheria toxin is extraordinarily potent; in sensitive species (e.g. humans, monkeys, rabbits, guinea pigs) as little as 100 to 150 ng/kg of body weight is lethal (Gill 1985). The relative sensitivity of a given sensitive eukaryotic cell line to the toxin has been shown to correlate with the number of cell surface receptors (Middlebrook *et al.* 1978). Since neither mouse nor rat cells express the diphtheria toxin receptor, they are resistant to its action (Pappenheimer 1977).

Immunity to diphtheria involves an antibody response to diphtheria toxin following clinical disease or immunization with diphtheria toxoid. Subclinical infection is no longer a source of diphtheria toxin antigen exposure and, if not boosted, antitoxin immunity wanes. As a conse-

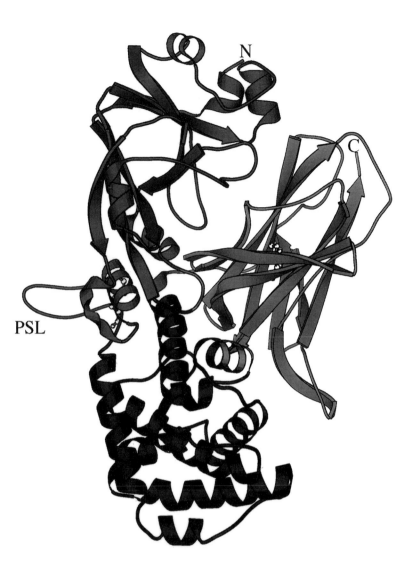

N

PSL

C

Figure 1. Ribbon diagram of the X-ray crystal structure of native diphtheria toxin (Choe *et al.* 1992) as modified by Bennett *et al.* (1994). The catalytic domain (*upper left*), transmembrane domain (*bottom*), and receptor binding domain (*right*) are shown. N, N-terminal end of the toxin; PSL, 14 amino acid protease sensitive loop which separates the catalytic from transmembrane and receptor binding domains; C, C-terminal end of the toxin. The ribbon diagram was generated using MOLESCRIPT (Kraulis 1991).

quence, a large percentage of the adults (30 to 60 per cent) currently have antitoxin levels that are below the protective level (0. 01 IU/ml) and are at risk. The adult population should be re-immunized with diphtheria toxoid every 10 years. Given the current diphtheria epidemic in Russia and other countries of the former Soviet Union, booster immunization with diphtheria–tetanus toxoids should be administered to all persons travelling to regions with high rates of endemic diphtheria (Central and South America, Africa, Asia, Russia, and Eastern Europe).

■ Use of diphtheria toxin-related fusion proteins in cell biology and clinical medicine

In recent years, the diphtheria toxin structural platform has been used in the genetic construction of a family of eukaryotic cell receptor-specific fusion proteins, or fusion toxins (Murphy *et al.* 1995). In each instance, the substitution of the native receptor binding domain with either a polypeptide hormone or growth factor (e. g. α-melanocyte stimulating hormone [α-MSH], epidermal growth factor [EGF], interleukin (IL)-2, IL-6, etc.) has resulted in the formation of a unique fusion toxin. These chimeric toxins, combine the receptor binding specificity of the cytokine with the cytotoxic potency of the transmembrane and catalytic domains of diphtheria toxin. In each instance, the fusion toxins have been shown to selectively intoxicate only those cells which bear the appropriate targeted receptor. Indeed, the first of these genetically engineered fusion toxins, DAB_{389} IL-2, is currently being evaluated in human clinical trials for the treatment of refractory lymphomas and autoimmune diseases where cells with high affinity IL-2 receptors play a major role in pathogenesis. Administration of DAB_{389} IL-2 has been shown to be safe, well tolerated, and capable of inducing durable remission from disease in the absence of severe adverse effects. It is likely that the diphtheria toxin-based fusion toxins will be important new biological agents for the treatment of specific tumors or disorders in which specific cell surface receptors may be targeted.

■ References

Bennett, M. J., Choe, S., and Eisenberg, D. (1994). Domain swapping entangling alliances between proteins. *Proc. Natl. Acad. Sci. USA*, **91**, 3127–31.

Bishai, W. R. and Murphy, J. R. (1988). In *The bacteriophages* (ed. I. Calendar), pp. 683–724, Plenum Publishing Corp, New York.

Bodley, J. W., Johnson, V. G., Wilson, B. A., Blanke, S. R., Murphy, J. R., Pappenheimer, A. M., Jr., and Collier, R. J. (1990). Does diphtheria toxin have nuclease activity? *Science*, **250**, 832–8.

Buck, G. A., Cross, R. E., Wong, T. P., Lorea, J., and Groman, N. B. (1985). DNA relationships among some tox-bearing corynebacteriophages. *Infect. Immun.*, **49**, 679–84.

Choe, S., Bennett, M. J., Fujii, G., Curmi, P. M. G., Kantardjieff, K. A., Collier, R. J. *et al.* (1992). The crystal structure of diphtheria toxin. *Nature*, **357**, 216–22.

Drazin, R., Kandel, J., and Collier, R. J. (1971). Structure and activity of diphtheria toxin. II. Attack by trypsin at a specific site within the intact toxin molecule. *J. Biol. Chem.*, **246**, 1504–10.

Gill, D. M. (1985). Bacterial toxins: a table of lethal amounts. *Microbiol. Rev.*, **46**, 86–94.

Gill, D. M. and Dinius, L. L. (1971). Observations on the structure of diphtheria toxin *J. Biol. Chem.*, **246**, 1485–91.

Greenfield, L., Bjorn, M. J., Horn, G., Fong, D., Buck, G. A., Collier, R. J. *et al.* (1983). Nucleotide sequence of the structural gene for diphtheria toxin carried by corynebacteriophage beta. *Proc. Natl. Acad. Sci. USA*, **80**, 6853–7.

Kraulis, P. J. (1991). MOLSCRIPT: a program to produce both detailed and schematic plots of protein structures. *J. Appl. Crystallogr.*, **24**, 946–50.

Middlebrook, J. L., Dorland, R. B., and Leppla, S. H. (1978). Association of diphtheria toxin with Vero cells. Demonstration of a receptor. *J. Biol. Chem.*, **253**, 7325–30.

Murphy, J. R., vanderSpek, J. C., Lemichez, E., and Boquet, P. (1995). In *Bacterial toxins, virulence factors and disease* Vol. 8, (ed. J. Moss, M. Vaughan, B. Iglweski, and A. Tu), pp. 23–45, Marcel Dekker, New York.

Naglich, J. G., Metherall, J. E., Russell, D. W., and Eidels, L. (1992). Expression cloning of a diphtheria toxin receptor: identity with a heparin-binding EGF-like growth factor precursor. *Cell*, **69**, 1051–61.

Pappenheimer, A. M., Jr (1977). Diphtheria toxin. *Annu. Rev. Biochem.*, **46**, 69–94.

Pappenheimer, A. M., Jr, Uchida, T., and Harper, A. (1972). An immunological study of the diphtheria toxin molecule. *Immunochemistry*, **9**, 891–906.

Ratti, G., Rappuoli, R., and Giannini, G. (1983). The complete nucleotide sequence of the gene coding for diphtheria toxin in the corynephage ω(tox⁺) genome. *Nucleic Acids Res.*, **11**, 6589–95.

Smith, W. P., Tai, P. C., Murphy, J. R. and Davis, B. D. (1980). Precursor in cotranslational secretion of diphtheria toxin. *J. Bacteriol.*, **141**, 184–9.

Tao, X., Schiering, N., Zeng, H.-Y., Ringe, D., and Murphy, J. R. (1994). Iron, DtxR, and the regulation of diphtheria toxin expression. *Mol. Microbiol.*, **14**, 191–7.

■ *John R. Murphy:*
Evans Department of Clinical Research and Department of Medicine,
Boston University Medical Center Hospital,
Boston, MA, 02118-2393
USA

Pseudomonas aeruginosa exotoxin A

Exotoxin A from Pseudomonas aeruginosa *is a 66 kDa single chain protein that enters the cytosol and ADP-ribosylates elongation factor 2. The toxin thereby blocks protein synthesis and kills the cell. The toxin is used to form targeted cytotoxic agents by linking it to ligands that bind to defined cell surface molecules.*

Pseudomonas aeruginosa exotoxin A (PEA, MW 66 kDa, 613 amino acids, sequence accession number: PSEETA K01397) consists of three major domains (Allured *et al.* 1986), as indicated in Fig. 1. The N-terminal domain Ia (amino acids 1–252) binds to the α_2-macroglobulin receptor at the cell surface (Kounnas *et al.* 1992), domain II (amino acids 253–364) mediates translocation of itself and domain III to the cytosol, and the C-terminal domain III (amino acids 405–613) catalyses the transfer of ADP-ribose from NAD to the diphthamide residue of elongation factor 2, which is thereby inactivated (Iglewski and Kabat 1975; Carroll and Collier 1987; Wick *et al.* 1990*a*, *b*). Intoxication is initiated by binding of the toxin

to its receptor and endocytotic uptake of the complex (Morris and Saelilnger 1986). Then follows proteolytic cleavage at a furin sensitive site between Arg²⁷⁹ and Gly²⁸⁰ and reduction of the disulfide between Cys²⁶⁵ and Cys²⁸⁷ (Ogata *et al.* 1990, 1992). The 37 kDa fragment thus generated is translocated to the cytosol. The translocation may occur from the lumen of the endoplasmic reticulum as the toxin contains a C-terminal REDLK sequence which, after removal of the C-terminal lysine residue, may function as an endoplasmic reticulum retention signal (Chaudhary *et al.* 1990). Removal of this signal strongly reduces the toxicity. Toxicity is also strongly reduced when His⁴²⁶ or Glu⁵⁵³ in domain III are

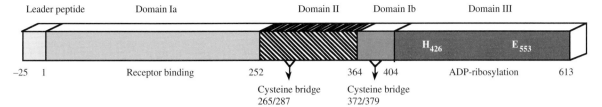

Figure 1. Schematic representation of *P. aeruginosa* exotoxin A.

mutated. In this case the toxin loses its enzymatic activity (Lukač and Collier 1988).

Domain Ia can be replaced by the variable domain of antibodies, or by lymphokines, growth factors and other molecules binding to defined cell surface molecules to make immunotoxins or related constructs for targeted cytotoxic therapy (Pastan and FitzGerald 1989).

Purification and sources

The toxin can be purified from the cell-free culture fluid of *Pseudomonas aeruginosa* (strain PA103, which produces little extracellular protease) as described (Iglewski and Sadoff 1979; Lory and Collier 1980). Wild-type and mutated toxin can also be purified from *E. coli* after transfection with the appropriate constructs (Gray *et al.* 1984; Benhar *et al.* 1994).

Toxicity

Toxicity can be assayed in a mouse lethality bioassay or in a cell toxicity assay (Iglewski and Sadoff 1979). The latter is fast and cheap and is most used. LD_{50} in mice weighing 17–20 g. is 0.1–0.3 μg. Intoxicated mice die within 2–6 days. The enzymatic activity can be assayed in an ADP-ribosylation assay (Iglewski and Sadoff 1979).

The toxin should be handled with care and one must avoid contact with the eyes or with open wounds. It is preferable to work with toxin solutions of 1 mg/ml or less. Lyophilized toxin should be handled with particular care to avoid inhalation.

Use in cell biology

Uncleaved PEA is not toxic to cells unless the cells are able to cleave the toxin, which occurs in the cells by furin (Moehring *et al.* 1993). PEA has been successfully used to select mutant cell lines lacking furin. The cells recover their toxin sensitivity upon transfection with furin or with the related Kex2 protease from *Saccharomyces cerevisiae*.

Use in targeted cytotoxic therapy

Due to its high toxic activity PEA has been extensively used in formation of immunotoxins and related constructs. The most common approach is to remove domain Ia and replace it with a ligand binding to cell surface molecules (Kondo *et al.* 1988). This may be the variable domain of immunoglobulins or various growth factors, hormones, and lymphokines. The toxicity of the constructs vary extensively. Endocytic uptake and transport to a deep membrane bounded compartment appears to be a prerequisite for translocation to the cytosol. The presence of a sequence resembling the retention signal for the endoplasmic reticulum (KDEL) suggests that transport of the toxin back to this location does take place. The finding that removal of the signal strongly reduces the toxicity indicates that this transport is required for translocation (Chaudhary *et al.* 1990).

References

Allured, V. S., Collier, R. J., Carrol, S. F., and McKay, D. B. (1986). Structure of exotoxin A of *Pseudomonas aeruginosa* at 3.0-Angstrom resolution. *Proc. Natl. Acad. Sci. USA*, **83**, 1320–4.

Benhar, I., Wang, Q.-c., FitzGerald, D., and Pastan, I. (1994). Pseudomonas exotoxin A mutants. Replacement of surface-exposed residues in domain III with cysteine residues that can be modified with polyethylene glycol in a site-specific manner. *J. Biol. Chem.*, **269**, 13398–404.

Carroll, S. F. and Collier, R. J. (1987). Active site of *Pseudomonas aeruginosa* exotoxin A. Glutamic acid 553 is photolabeled by NAD and shows functional homology with glutamic acid 148 of diphtheria toxin. *J. Biol. Chem.*, **262**, 8707–11.

Chaudhary, V. K., Jinno, Y., FitzGerald, D. J., and Pastan, I. (1990). Pseudomonas exotoxin contains a specific sequence at the carboxyl terminus that is required for cytotoxicity. *Proc. Natl. Acad. Sci. USA*, **87**, 308–12.

Gray, G. L., Smith, D. H., Baldridge, J. S., Harkins, R. N., Vasil, M. L., Chen, E. Y. *et al.* (1984). Cloning, nucleotide sequence, and expression in *Escherichia coli* of the exotoxin A strucutral gene of *Pseudomonas aeruginosa*. *Proc. Natl. Acad. Sci. USA*, **81**, 2645–9.

Iglewski, B. H. and Kabat, D. (1975). NAD-dependent inhibition of protein synthesis by *Pseudomonas aeruginosa* toxin. *Proc. Natl. Acad. Sci. USA*, **72**, 2284–8.

Iglewski, B. H. and Sadoff, J. C. (1979). Toxin inhibitors of protein synthesis: production, purification, and assay of *Pseudomonas aeruginosa* toxin A. *Methods Enzymol.*, **60**, 780–93.

Kondo, T., FitzGerald, D., Chaudhary, V. K., Adhya, S., and Pastan, I. (1988). Activity of immunotoxins constructed with modified Pseudomonas exotoxin A lacking the cell recognition domain. *J. Biol. Chem.*, **263**, 9470–5.

Kounnas, M. Z., Morris, R. E., Thompson, M. R., FitzGerald, D. J., Strickland, D. K., and Saelinger, C. B. (1992). The alpha 2-macroglobulin receptor/low density lipoprotein receptor-related protein binds and internalizes Pseudomonas exotoxin A. *J. Biol. Chem.*, **267**, 12420–3.

Lory, S. and Collier, R. J. (1980). Expression of enzymic activity by exotoxin A from *Pseudomonas aeruginosa*. *Infect. Immun.*, **28**, 494–501.

Lukač, M. and Collier, R. J. (1988). Restoration of enzymic activity and cytotoxicity of mutant, E553C, *Pseudomonas aeruginosa* exotoxin A by reaction with iodoacetic acid. *J. Biol. Chem.*, **263**, 6146–9.

Moehring, J. M., Inocencio, N. M., Robertson, B. J., and Moehring, T. J. (1993). Expression of mouse furin in a Chinese hamster cell resistant to Pseudomonas exotoxin A and viruses complements the genetic lesion. *J. Biol. Chem.*, **268**, 2590–4.

Morris, R. E. and Saelinger, C. B. (1986). Reduced temperature alters Pseudomonas exotoxin A entry into the mouse LM cell. *Infect. Immun.*, **52**, 445–53.

Ogata, M., Chaudhary, V. K., Pastan, I., and FitzGerald, D. J. (1990). Processing of Pseudomonas exotoxin by a cellular protease results in the generation of a 37,000-Da toxin fragment that is translocated to the cytosol. *J. Biol. Chem.*, **265**, 20678–85.

Ogata, M., Fryling, C. M., Pastan, I., and FitzGerald, D. (1992). Cell-mediated cleavage of Pseudomonas exotoxin between Arg279 and Gly280 generates the enzymatically active fragment which translocates to the cytosol. *J. Biol. Chem.*, **267**, 25396–401.

Pastan, I. and FitzGerald, D. (1989). Pseudomonas exotoxin: chimeric toxins. *J. Biol. Chem.*, **264**, 15157–60.

Wick, M. J., Hamond, A. N., and Iglewski, B. H. (1990a). Analysis of the structure–function relationship of *Pseudomonas aeruginosa* exotoxin A. *Mol. Microbiol.*, **4**, 527–35.

Wick, M. J., Frank, D. W., Storey, D. G., and Iglewski, B. H. (1990b). Structure, function, and regulation of *Pseudomonas aeruginosa* exotoxin A. *Annu. Rev. Microbiol.*, **44**, 335–63.

■ *Sjur Olsnes:*
Department of Biochemistry,
Institute for Cancer Research,
The Norwegian Radium Hospital,
Montebello, 0310 Oslo, Norway

Shiga toxins (*Shigella dysenteriae* serotype I, *Escherichia coli*)

Shiga toxins are potent bacterial protein toxins that inhibit protein synthesis in target eukaryotic cells by cleaving an adenine residue from cytoplasmic ribosomes such that the ribosome no longer interacts with elongation factors eEF-1 and eEf-2.

The Shiga toxins (also known as verotoxins) comprise a family of toxins that inhibit protein synthesis in intoxicated cells (for reviews see O'Brien and Holmes 1987; O'Brien *et al.* 1992; Obrig 1994). The holotoxin consists of an A polypeptide (M_r 32 000) that has *N*-glycosidase activity which removes a single adenine residue from the 28 S rRNA (the same mode of action is found in the plant toxin ricin), and a pentamer of B polypeptides (M_r 7700 each) that confers specificity for the eukaryotic cell receptor, globotriaosylceramide or Gb$_3$ (Lingwood 1993). Toxin entry into the eukaryotic cell occurs via receptor-mediated endocytosis with subsequent routing of the toxin from the transgolgi through the endoplasmic reticulum to the cytosol (Sandvig and Deurs 1994). The A subunit contains a trypsin cleavage site near the C-terminus, but the importance of this nicking site for cytotoxicity is still under investigation. The cleaved A subunit is held together by a disulfide bridge that may get reduced after the toxin has been internalized. The crystal structure of Shiga toxin has been solved to 2.5 Å (Stein *et al.* 1992; Fraser *et al.* 1994). It is organized in two different domains: a pentameric oligosaccharide binding domain and a catalytic domain. The active site of the toxin appeared to be blocked by a portion of the A subunit, a finding which indicates that a conformational change may be required for full toxin activity. The active site of Shiga toxin appears to be the glutamic acid residue at position 167 in the mature A subunit, although neighbouring amino acids also appear to play a role (O'Brien *et al.* 1992).

The Shiga toxin family consists of Shiga toxin from *S. dysenteriae* and Shiga toxin types 1 and 2 found in *E. coli*. Shiga toxin and Stx1 are essentially identical, with only one amino acid difference in the mature A subunit. Shiga toxin is about 60 per cent homologous to Stx2 at the amino acid level but polyclonal antisera to one toxin type does not cross react with the heterologous toxin. Variants of Shiga toxin type 2 are recognized based on biological differences or divergence in immunological reactivity or receptor binding. Based on these criteria, two variants of Stx2 have been identified: Stx2c (VT2c), which can be distinguished antigenically from Stx2 (Schmitt et al. 1991), and the edema disease toxin, Stx2e (VT2e, formerly called Stx2v or VT2v), which preferentially uses globotetraosylceramide (Gb$_4$) as a functional receptor rather than Gb$_3$ (Lingwood 1993).

Shiga toxin is chromosomally encoded (GenEMBL accession number M19437). Stxs are also found on the chromosome (Stx2e, some Stx1) or on bacteriophages (Stx1 and Stx2 GenEMBL accession numbers are M16625 and X07865, respectively). However, the toxins do share a conserved genetic structure in which the A and B subunit are encoded in an operon (O'Brien et al. 1992). The expression of Shiga toxin and Stx1, but not Stx2, is environmentally regulated such that high levels are made under iron-restricted conditions (O'Brien and Holmes 1987; O'Brien et al. 1992).

Shiga toxin producing organisms cause dysentery (S. dysenteriae), hemorrhagic colitis (E. coli), and the hemolytic uremic syndrome (S. dysenteriae, E. coli). One model for the role of these toxins in pathogenesis of disease is that the toxin binds to mature columnar epithelial cells in the intestine of the infected human or animal, and the toxin then enters those cells and halts protein synthesis. This inhibition of protein synthesis, in turn, leads to cell death and a subsequent loss of intestinal absorptive capacity (O'Brien and Holmes 1987). Recent hypotheses for how more serious Shiga toxin mediated sequelae may arise involve systemic delivery of toxin from the intestine followed by binding of the toxin to receptors on vascular endothelial cells of the central nervous system (edema disease), colon (hemorrhagic colitis), and kidney (HUS). Toxin bound to glomerular endothelia may cause direct damage to those cells or may act in concert with cytokines to mediate further damage to the endothelium (Tesh and O'Brien 1991).

■ Purification and sources

Shiga toxin can be isolated from shigellae grown in low iron medium or from K-12 strains carrying toxin clones or toxin-encoding phage. These toxin-producing bacteria are lysed, and toxin is purified by a series of step that include ammonium sulfate precipitation, anion exchange chromatography, chromatofocusing, and finally antitoxin affinity chromatography (O'Brien and Holmes 1987; O'Brien et al. 1992; Lindgren et al. 1994).

■ Toxicity

The intravenous or intraperitoneal LD$_{50}$ for Shiga and Shiga toxin type 1 in mice has been reported to be from 28 ng/mouse (Yutsudo et al. 1986) to 400 ng/mouse (0.02 mg/kg; Tesh et al. 1993) to 0.450 mg/kg, (Eiklid and Olsnes 1983). For Shiga toxin type 2, the mouse LD$_{50}$ is approximately 1 ng/mouse (5×10^{-5} mg/kg; Tesh et al. 1993; Lindgren et al. 1994). The intravenous LD$_{50}$ of Shiga toxin for the rabbit is about 0.2 μg/kg (Richardson et al. 1992).

■ Use in cell biology

Shiga toxin has been used to study the endocytic pathway. In cells treated with butyric acid, Sandvig et al. (1992) have demonstrated that this toxin enters endosomal compartments, wherefrom it reaches the Golgi apparatus and the endoplasmic reticulum. Thus, the study of the membrane trafficking of this toxin revealed a pathway from the plasma membrane up to endosplasmic reticulum, that may be followed by more physiological ligands. It is expected that Shiga toxin will be a very useful marker, particular in sudies involving polarized and neuronal cells.

References

Eiklid, K. and Olsnes, S. (1983). Animal toxicity of *Shigella dysenteriae* cytotoxin: Evidence that the neurotoxic, enterotoxic, and cytotoxic activities are due to one toxin. *J. Immunol.*, **130**, 380–4.

Fraser, M., Chernaia, M., Kozlov, Y., and James, M. (1994). Crystal structure of the holotoxin from *Shigella dysenteriae* at 2.5 Å resolution. *Struct. Biol.*, **1**, 59–64.

Lindgren, S., Samuel, J., Schmitt, C., and O'Brien, A. (1994). The specific activities of Shiga-like toxin type II (SLT-II) and SLT-II-related toxins of enterohemorrhagic *Escherichia coli* differ when measured by vero cell cytotoxicity but not by mouse lethality. *Infect. Immun.*, **62**, 623–31.

Lingwood, C. (1993). Verotoxins and their glycolipid receptors. *Adv. Lipid Res.*, **25**, 189–211.

O'Brien, A. D. and Holmes, R. K. (1987). Shiga and Shiga-like toxins. *Microbiol. Rev.*, **51**, 206–20.

O'Brien, A., Tesh, V., Donohue-Rolfe, A., Jackson, M., Olsnes, S., Sandvig, K., *et al.* (1992). Shiga toxin: Biochemistry, genetics, mode of action, and role in pathogenesis. *Curr. Top. Microbiol. Immunol.*, **180**, 65–94.

Obrig, T. (1994). Toxins that inhibit host protein synthesis. *Methods Enzymol.*, **235**, 647–56.

Richardson, S., Rotman, T., Jay, V., Smith, C., Becker, L., Petric, M., *et al.* (1992). Experimental verocytotoxemia in rabbits. *Infect. Immun.*, **60**, 4154–67.

Sandvig, K. and Deurs, B. (1994). Endocytosis and intracellular sorting of ricin and Shiga toxin. *FEBS Lett.*, **346**, 99–102.

Sandvig, K., Garred, O., Prydz, J, Kozlov, J. V., Hansen, H. and van Deurs, B. (1992). Retrograde transport of endocytosed Shiga toxin to the endoplasmic reticulum. *Nature*, **358**, 510–2.

Schmitt, C., McKee, M., and O'Brien, A. (1991). Two copies of Shiga-like toxin II-related genes common in enterohemorrhagic *Escherichia coli* strains are responsible for the antigenic heterogeneity of the O157:H_ strain E32511. *Infect. Immun.*, **59**, 1065–73.

Stein, P. E., Boodhoo, A., Tyrrell, G. J., Brunton, J. L., and Read, R. J. (1992). Crystal strucutre of the cell-binding B oligomer of verotoxin-1 from *E. coli. Nature*, **355**, 748–50.

Tesh, V. and O'Brien, A. (1991). The pathogenic mechanisms of Shiga toxin and the Shiga-like toxins. *Mol. Microbiol.*, **5**, 1817–22.

Tesh, V., Burris, J., Owens, J., Gordon, V., Wadolkowski, E., O'Brien, A. *et al.* (1993). Comparison of the relative toxicities of Shiga-like toxins type I and type II for mice. *Infect. Immun.*, **61**, 3392–402.

Yutsudo, T., Honda, T., Miwatani, T., and Takeda, Y. (1986). Characterization of purified Shiga toxin from *Shigella dysenteriae* 1. *Microbiol. Immunol.*, **30**, 1115–27.

■ *Angela R. Melton-Celsa and Alison D. O'Brien:*
Department of Microbiology and Immunology,
Uniformed Services University of the Health Sciences,
4301 Jones Bridge Road,
Bethesda, MD 20814,
USA

Ricin (*Ricinus communis*)

Ricin is a protein toxin found in the seeds of Ricinus communis. *The various isoforms of ricin have molecular weights of 63–66 kDa, and the toxin consists of two polypeptide chains connected by a disulfide bond. One of the chains, the B-chain binds to cell-surface glycoproteins and glycolipids with terminal galactose, whereas the other chain, the A-chain, enters the cytosol and blocks protein synthesis enzymatically by removal of one adenine from the 28 S RNA of the 60 S ribosomal subunit. Endocytosis and intracellular transport of ricin to the Golgi apparatus and possibly to the ER seem to be required for translocation of the A-chain to the cytosol.*

Ricin is synthesized in the endosperm cells of *Ricinus communis* seeds, and there are various isoforms with molecular weights of 63–66 kDa. The molecule is glycosylated (for review, see Barbieri *et al.* 1993; Lord *et al.* 1994). The ricin gene codes for a preprotoxin molecule which is processed in the plant (preproricin D: 576 amino acids, GenBank accession no. X52908). The leader sequence of ricin D has 35 amino acid residues, and the interchain linker between the two chains has 12 amino acids (Lord *et al.* 1994). The mature form of ricin D (Fig. 1) has an A-chain with 267 amino acids and a B-chain with 262 residues.

The intoxication consists of the following steps: The B-chain binds to cell surface receptors with terminal galactose and the toxin is endocytosed and transported to the organelle from where translocation of the A-chain to the cytosol takes place. Transport to the Golgi apparatus and possibly also retrograde transport to the endoplasmic reticulum (ER) may be involved in the internalization process (Pelham *et al.* 1992; Wales *et al.* 1992; Sandvig and van Deurs 1994*a*) (Fig. 2). After translocation to the cytosol, the A-chain inhibits the protein synthesis. Ricin A-chain is a *N*-glycosidase which removes one adenine from the 28 S RNA of the 60 S subunit of the ribosome (Endo *et al.* 1987). The toxin has no effect on prokaryotic ribosomes (Lord *et al.* 1994).

Ricin or parts of the toxin molecule, for instance the A-chain, is being used to construct immunotoxins or other conjugates where the aim is to selectively kill a subgroup of cells, for instance cancer cells (Olsnes *et al.* 1989; Barbieri *et al.* 1993; Lord *et al.* 1994).

Toxins with a similar two-chain structure as ricin are the plant proteins abrin, modeccin, volkensin, and viscumin

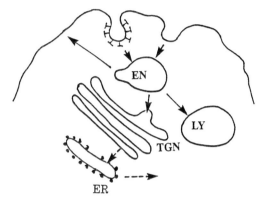

Figure 2. Intracellular pathways followed by ricin. Known pathways, arrows with intact lines; hypothetical pathways, arrows with broken lines. The toxin is endocytosed from clathrin-coated pits and also by a clathrin-independent mechanism, and it is transported to endosomes (EN), from where it may be recycled. The toxin is also transported to lysosomes (LY), to the trans-Golgi network (TGN), and possibly retrogradely through the Golgi to the endoplasmic reticulum (ER) from where it may be translocated to the cytosol.

which come from different plants, but which all inhibit protein synthesis in the same way as ricin. In spite of the similarities, the entry mechanisms of these toxins may differ (for review, see Olsnes *et al.* 1990).

■ Purification and sources

Ricin can be purified from the seeds of *Ricinus communis* (Olsnes 1978). The seeds are extracted with acetic acid, and ricin is purified by chromatography on CM-cellulose and Sepharose 4B. Ricin can also be purchased from Sigma Chemical Company, St. Louis, Mo., USA.

■ Toxicity

For humans the maximum tolerated dose has been estimated to be 23 μg/m^2 (about 40 μg for a person of 60 kg)

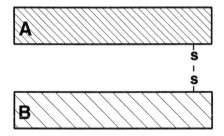

Figure 1. Schematic structure of ricin.

(Fodstad *et al.* 1984). LD_{50} for mice after parenteral injection is 2. 6 μg/kg (Barbieri *et al.* 1993), and the minimum lethal dose for dogs is 1. 75 μg/kg (Fodstad *et al.* 1984). The lethal dose of castor oil seed, the source of ricin, has been measured on a number of different animals (Balint 1974). The horse was found to be very sensitive, the lethal dose was 0. 1 g/kg, whereas the lethal dose for a hen is 14 g/kg. For safety, gloves should be used when working with ricin. If possible, avoid freeze-drying the toxin since this will increase the risk of inhaling the toxin.

■ Use in cell biology

Ricin has proven valuable in studies of endocytosis and intracellular trafficking (van Deurs *et al.* 1993; Eker *et al.* 1994; Sandvig and van Deurs 1994*a*). Since the toxin binds to a large number of different glycolipids and glycoproteins, ricin can be used as a membrane marker. Experiments with ricin have provided evidence for the existence of more than one endocytic pathway (Sandvig and van Deurs 1994*b*). The toxin is endocytosed even when the formation of coated vesicles from coated pits is blocked by acidification of the cytosol, and also when clathrin-coated pits are removed by potassium depletion of cells. Furthermore, ricin is transported to the lysosomes and to the Golgi apparatus, it is transcytosed across epithelial cell layers and it is recycled, and the toxin can therefore be used to study all these pathways. The mechanism for ricin translocation to the cytosol is of importance also in connection with the construction of immunotoxins, which often contain parts of the native ricin molecule (Olsnes *et al.* 1989; Barbieri *et al.* 1993; Lord *et al.* 1994). The efficiency of hybrid molecules like immunotoxins could be dependent on routing to the correct intracellular compartment for translocation to the cytosol. There is evidence that ricin has to be transported to the Golgi apparatus before translocation to the cytosol (for review, see Sandvig and van Deurs 1994*a*), and retrograde transport to the ER may be required for translocation to occur (Pelham *et al.* 1992; Wales *et al.* 1992; Sandvig and van Deurs 1994*a*). Ricin does not contain the ER retrieval signal KDEL, and more knowledge about the possible retrograde transport of this molecule should therefore provide important new information about trafficking of proteins in the Golgi/ER system. Some of the known and hypothetical pathways followed by ricin are shown in Fig. 2.

■ References

Balint, G. A. (1974). Ricin: The toxic protein of castor oil seeds. *Toxicology*, **2**, 77–102.

Barbieri, L., Battelli, M. G., and Stirpe, F. (1993). Ribosome-inactivating proteins from plants. *Biochim. Biophys. Acta*, **1154**, 237–82.

Eker, P., Holm, P. K., van Deurs, B., and Sandvig, K. (1994). Selective regulation of apical endocytosis in polarized MDCK cells by mastoparan and cAMP. *J. Biol. Chem.*, **269**, 18607–15.

Endo, Y., Mitsui, K., Motizuki, M., and Tsurugi, K. (1987). The mechanism of action of ricin and related toxic lectins on eukaryotic ribosomes. The site and the characteristics of the modification in the 28 S ribosomal RNA caused by the toxins. *J. Biol. Chem.*, **262**, 5908–12.

Fodstad, Ø., Kvalheim, G., Godal, A., Lotsberg, J., Aamdal, S., Høst, H., *et al.* (1984). Phase I study of the plant protein ricin. *Cancer Res.*, **44**, 862–5.

Lord, J. M., Roberts, L. M. and Robertus, J. D. (1994). Ricin: structure, mode of action, and some current applications. *FASEB J.*, **8**, 201–8.

Olsnes, S. (1978). The toxic lectins ricin and ricinus agglutinin. *Methods Enzymol.*, **50**, 330–5.

Olsnes, S., Sandvig, K., Petersen, O. W., and van Deurs, B. (1989). Immunotoxins—entry into cells and mechanism of action. *Immunol. Today*, **10**, 291–5.

Olsnes, S., Stenmark, H., Øivind Moskaug, J., McGill, S., Madshus, I. H. and Sandvig, K. (1990). Protein toxins with intracellular targets. *Microb. Pathog.*, **8**, 163–8.

Pelham, H. R. B., Roberts, L. M., and Lord, J. M. (1992). Toxin entry: how reversible is the secretory pathway? *Trends Cell Biol.*, **2**, 183–5.

Sandvig, K. and van Deurs, B. (1994*a*). Endocytosis and intracellular sorting of ricin and Shiga toxin. *FEBS Lett.*, **346**, 99–102.

Sandvig, K. and van Deurs, B. (1994*b*). Endocytosis without clathrin. *Trends Cell. Biol.*, **4**, 275–7.

van Deurs, B., Hansen, S. H., Olsnes, S., and Sandvig, K. (1993). Protein uptake and cytoplasmic access in animal cells. In *Biological barriers to protein delivery* (ed. K. L. Audus and T. J. Raub), pp. 71–104, Plenum Press, New York.

Wales, R., Chaddock, J. A., Roberts, L. M., and Lord, J. M. (1992). Addition of an ER retention signal to the ricin A chain increases the cytotoxicity of the holotoxin. *Exp. Cell Res.*, **203**, 1–4.

■ *Kirsten Sandvig:*
Department of Biochemistry,
Institute for Cancer Research,
The Norwegian Radium Hospital,
Montebello, N-0310 Oslo,
Norway

Ribosome-inactivating proteins

Single-chain (type 1, as opposed to type 2, two-chain proteins) ribosome inactivating proteins are plant proteins possessing N-glycosidase activity which release adenine from rRNA, thus rendering ribosomes unable to perform proteins synthesis. They have antiviral properties and can be linked to antibodies from immunotoxins, selectively toxic to immunotoxins, selectively toxic to the cells recognized by the antibody.

Ribosome-inactivating proteins (RIPs, review by Barbieri *et al.* 1993) are a class of proteins present in various tissues of several plants which inactivate mammalian ribosomes and, with less activity and to variable extent, plant, fungal, and bacterial ribosomes. They are enzymes, *N*-glycosidases, which release adenine from rRNA. They are divided into two groups: type 1, consisting of a single peptide chain, and type 2, in which an A chain with enzymatic activity is linked to a slightly larger B chain with the properties of a galactosyl-specific lectin. The B chain binds to galactosyl-terminated receptors on the membrane of most animal cells, allowing the molecule to enter the cytoplasm. Thus most type 2 RIPs are potent toxins, ricin being the best known, although recently some nontoxic lectins with the structure of type 2 RIPs have been described. Type 1 RIPs do not enter easily into cells, and consequently are much less toxic than type 2 RIPs.

Sources, purification, and properties

Type 1 RIPs are widely distributed in the plant kingdom (Barbieri *et al.* 1993) and can be found in several tissues (roots, stems, leaves, seeds, latex) of many plants. Their concentration in plant tissues is highly variable, being particularly high in plants belonging to some families (Caryophyllaceae, Phytolaccaceae, Euphorbiaceae, Cucurbitaceae).

They can be purified by ion-exchange chromatography (Barbieri *et al.* 1987), taking advantage of their high isoelectric point. In some cases, additional chromatography on Blue Sepharose or gel filtration may be required.

Type 1 RIPs are 30 kDa proteins. Most of them are glycoproteins, and have a p*I* in a strongly basic region, usually > 9.5. A number of them have been sequenced and cloned.

Mechanism of action

Ribosome-inactivating proteins are *N*-glycosidases, and damage eukaryotic ribosomes by breaking the N-glycosidic bond of adenine in a precise position of rRNA (A_{4324} in the case of rat liver ribosomes), in a GAGA sequence in a loop at the top of a stem (review by Endo 1988). This renders ribosomes unable to bind elongation factor 2 (in same cases also elongation factor 1), thus arresting protein synthesis. RIPs act also on bacterial ribosomes and on purified rRNA, although at concentrations much higher than those effective on mammalian ribosomes. Some RIPs remove more than one adenine from all types of RNA tested, as well as from poly(A) and from DNA (Barbieri *et al.* 1994).

Toxicity and cytotoxicity

Type 1 RIPs do not have the B chain with lectin properties of type 2 RIPs, and consequently do not bind to cells and enter less easily into them. Thus they are much less toxic to cells and animals than type 2 RIPs: their LD_{50} for mice is of the order of mg/kg of body weight. They are very potent inhibitors of cell-free protein synthesis, with IC_{50}s (concentration causing 50 per cent inhibition of protein synthesis by HeLa cells) in the range 0.2–9.2 μM. Some cells (macrophages, trophoblasts, some malignant cell lines) are more sensitive to RIPs.

Biological properties

Single-chain ribosome-inactivating proteins have antiviral activity against both plant and animal viruses. It is commonly accepted that they act by entering more easily into cells infected by viruses, thus inactivating ribosomes and killing the cells, with consequent arrest of viral proliferation. They have also immunosuppressive activity, possibly due to killing of macrophages, for which they have a higher toxicity.

Possible uses

Ribosome-inactivating proteins can be linked to antibodies or other appropriate carriers (growth factors, hormones) to make immunotoxins or other conjugates selectively toxic to the cells recognized by the carrier (Frankel 1988; Magerstädt 1991). Immunotoxins can be used to selectively eliminate a given type of cells *in vitro* and *in vivo*, and are being experimented

(1) to remove contaminant cells from cultures;

(2) to remove malignant or immunocompetent cells *ex vivo* from bone marrow to be transplanted; and

(3) in clinical trials to treat patients with malignant tumours (Vallera 1994; Siegall *et al.* 1995).

Immunotoxins and other conjugates are used also to cause selective lesions in the central nervous system (Wiley and Lappi 1994).

Plants have been transfected with RIP genes to confer them with resistance to viruses (Lodge *et al.* 1993) or fungi (Logemann *et al.* 1992).

■ References

Barbieri, L., Stoppa, C., and Bolognesi, A. (1987). Large scale chromatographic purification of ribosome-inactivating proteins. *J. Chromatogr.*, **408**, 235–43.

Barbieri, L., Battelli, M. G. and Stirpe, F. (1993). Ribosome-inactivating proteins from plants. *Biochim. Biophys. Acta*, **1154**, 237–82.

Barbieri, L., Gorini, P., Valbonesi, P., Castiglioni, P., and Stirpe, F. (1994). Unexpected activity of saporins. *Nature*, **372**, 624.

Endo, Y. (1988). Mechanism of action of ricin and related toxins on the inactivation of eukaryotic ribosomes. In: *Immunotoxins* (ed. A. E. Frankel), pp. 75–89, Kluwer Academic, Boston.

Frankel, A. E. (ed.) (1988). *Immunotoxins,* Kluwer Academic, Boston.

Lodge, J. K., Kaniewski, W. K., and Tumer, N. E. (1993). Broad-spectrum virus resistance in transgenic plants expressing poke-weed antiviral protein. *Proc. Natl. Acad. Sci. USA*, **90**, 7089–93.

Logemann, J., Jach, G., Tommerup, H., Mundy, J., and Schell, J. (1992). Expression of a barley ribosome-inactivating protein leads to increased fungal protection in transgenic tobacco plants. *Bio/Technol.*, **10**, 305–8.

Magerstädt, M. (1991). Immunotoxins. In *Antibody conjugates and malignant disease*, pp. 151–93, CRC Press, Boca Raton, Florida.

Siegall, C. B., Wolff, E. A., Gawlak, S. L., Paul, L., Chace, D., and Mixan, B. (1995). Immunotoxins as cancer chemotherapeutic agents. *Drug Develop. Res.*, **34**, 210–19.

Vallera, D. A. (1994). Immunotoxins: will their clinical promise be filled? *Blood*, **83**, 309–17.

Wiley, R. G. and Lappi, D. A. (1994). *Suicide transport and immunolesioning*, R. G. Landes Company, Austin.

■ *Fiorenzo Stirpe:*
Dipartimento di Patologia Sperimentale,
Università di Bologna,
Via S. Giacomo 14,
41026 Bologna, Italy

α-Sarcin and related toxins (*Aspergillus*)

α-Sarcin is a nonglycosylated single polypeptide chain of 17 kDa secreted by the mould Aspergillus giganteus. *It binds cell membrane and enters the cytosol where it specifically cleaves a single phosphodiester bond of the 28S rRNA, that causes ribosome disruption and impairs protein biosynthesis. This protein belongs to a family of fungal cytotoxic ribonucleases composed of mitogillin (produced by* A. restrictus*), restrictocin (from* A. restrictus*) and Asp fI (from* A. fumigatus*).*

Aspergillins are a family of toxic extracellular fungal ribonucleases displaying a high amino acid sequence homology (Table 1). α-Sarcin, mitogillin, and restrictocin were discovered and characterized as antitumoral agents displaying *in vivo* cytotoxicity (Olson and Goerner 1965; Goldin *et al.* 1966). However, their clinical use was hampered since they revealed to be too toxic (Roga *et al.* 1971). Asp fI was recently discovered as an IgE-binding protein involved in the pathogenesis of invasive aspergillosis (Arruda *et al.* 1990; Lamy *et al.* 1991).

Aspergillins are encoded by a single gene that is translated into a precursor with a 27 residue amino terminal leader sequence that is removed upon maturation (Oka *et al.* 1990; Lamy and Davies 1991; Moser *et al.* 1992). The mechanism by which host ribosomes are protected from the toxin catalysis has not been deciphered yet although

Table 1 Aspergillins

Aspergillin	Source	Amino acid residues	Swiss Prot. Data Bank accession number	Gen Bank accession number
α-Sarcin	*A. giganteus* MDH 18894	150	P00655	X53394 (*sar*)
Mitogillin	*A. restrictus* NRRL 3050	149	PO4389 P19792	M94249 (*mif*)
Restrictocin	*A. restrictus* ATCC 34475, NRLL 2869	149	PO4389 P19792	M65257 (*res*)
Asp fI	*A. fumigatus* ATCC 42202, CBS 143.89	149	PO4389 P19792	M83781 (*asp fla*)

several hypotheses have been proposed (Lamy and Davies 1991). The mature form consists of a highly polar single polypeptide chain of 149 amino acid residues (150 in α-sarcin) that contains two intramolecular disulfide bonds and is rich in basic residues (Martínez del Pozo et al. 1988; Mancheño et al. 1995). The protein folds into a single $\alpha\beta$ domain that exhibits a highly specific ribonucleolytic activity and the ability to interact with acid phospholipids (Gasset et al. 1994). Active site residues have been identified based on sequence similarities with other nontoxic phylogenetically related ribonucleases (Martínez del Pozo et al. 1988; Mancheño et al. 1995). Positive-charged loops and a theoretically predicted hydrophobic antiparallel β-sheet structurally support the electrostatic and hydrophobic interactions with membranes (Mancheño et al. 1995).

■ Purification and sources

α-Sarcin, mitogillin, restrictocin, and Asp fI are isolated and purified to homogeneity from culture supernatants of the respective *Aspergillus* genus producing strains by two chromatographic steps (Arruda et al. 1990; Gasset et al. 1994 and references therein). Several expression systems with suitable purification protocols for recombinant wild type and mutant protein forms are also available for all members of the family (Lamy and Davies 1991; Better et al. 1992; Moser et al. 1992; Lacadena et al. 1994). Purity assessment by SDS-PAGE and activity determination are recommended. Specific ribonuclease activity is ascertained by the appearance of α-fragment from the 28S rRNA of eukaryotic ribosomes using cell-free rabbit reticulocyte lysates (Endo and Wool 1982).

■ Toxicity

α-Sarcin *in-vivo* toxicity is tested by intraperitoneal injection of different amounts of toxin in Balb/c mice (LD_{50} = 12 mg/kg) (Olmo and Lizarbe, in preparation). Mitogillin LD_{50} was determined following a 24 i.v. daily injection protocol (LD_{50} = 0.17 mg/kg/day; dog LD_{50} = 0.035 mg/kg/day; monkey LD_{50} = 0.06 mg/kg/day) (Roga et al. 1971). Concentrations below the LD_{50} successfully inhibit the growth of different animal tumours (Roga et al. 1971; Olmo and Lizarbe, in preparation). The use of these toxins does not present any problem for healthy operators. The recent incorporation of Asp f1 to the aspergillin family underlines their potential harmful behaviour as allergens (Arruda et al. 1990; Lamy et al. 1991).

■ Use in cell biology

The cytotoxic action exhibited by these proteins is due to their ability to enter cells and hydrolyse a single phosphodiester bond on the 3′side of G4325 in eukaryotic 28S RNA, that results in protein biosynthesis inhibition (Endo and Wool 1982; Wool 1984). This bond is located in a universal conserved region, known as the α-sarcin loop,

involved in the EF-1-dependent binding of aminoacyl-tRNA and EF-2-catalysed GTP hydrolysis and consequently essential for ribosome function (Wool et al. 1992). This highly specific ribonuclease actvity has been very useful in determining ribosome structure and protein biosynthesis mechanism. The reason why these proteins are not sensitive to the action of intracellular ribonuclease inhibitors is still an open question.

Aspergillins were widely used to probe cell membrane permeability to macromolecules since no surface receptor was found (Alonso and Carrasco 1980; Fernández-Puentes and Carrasco 1980; Otero and Carrasco 1986, 1988). Notwithstanding, aspergillins enter mammalian transformed cells interfering the proliferation pattern, inhibiting protein biosynthesis, and leading to cell death (Moser et al. 1992; Olmo et al. 1993; Turnay et al. 1993). The IC_{50} for protein biosynthesis varies from 0.01–0.4 nM in cell-free systems to 0.3–10 μM in intact cells (Fando et al. 1985; Turnay et al. 1993). These serial events emphasize the potential use of aspergillins as apoptosis-inducers. Furthermore, incorporation of aspergillin encoding genes into suicide cassettes, under the control of highly regulated promoter, will be useful to establish the function of specific cells and tissues during development. The cytotoxic efficiency is dramatically enhanced when these proteins are covalently linked to anti-cell specific antibodies (Conde et al. 1989; Wawrzynczak et al. 1991; Better et al. 1992). Fluid phase endocytosis is probably the pathway into the cytosol (Olmo et al. 1993). Biophysical studies have shown that α-sarcin binds, inserts, and translocates across acid phospholipid-containing model membranes (Gasset et al. 1994).

■ References

Alonso M. A. and Carrasco, L. (1980). Permeabilization of mammalian cells to proteins by the ionophore nigericin. *FEBS Lett.*, **127**, 112–14.

Arruda, K. L., Platts-Mills, T. A. E., Fox, J. W., and Chapman, M. (1990). Aspergillus fumigatus allergen I, a major IgE-binding protein, is a member of the mitogillin family of cytotoxins. *Exp. Med.*, **172**, 1529–32.

Better, M., Bernhard, S. L., Lei, S.-P., Fishwild, D. M., and Carroll, S. F. (1992). Activity of the recombinant mitogillin and mitogillin immunoconjugates. *J. Biol. Chem.*, **267**, 16712–18.

Conde, F. P., Orlandi, R., Canevari, S., Mezzanzanica, D., Ripamonti, M., Muñoz, S. M., et al. (1989). The Aspergillus toxin restrictocin is a suitable cytotoxic agent for generation of immunoconjugates with monoclonal antibodies directed against human carcinoma cells. *Eur. J. Biochem.*, **178**, 795–802.

Endo, Y. and Wool, I. G. (1982). The site of action of α-sarcin on eukaryotic ribosomes. *J. Biol. Chem.*, **257**, 9054–60.

Fando, J. L., Alaba, I., Escarmis, C., Fernández-Luna, J. L., Mendez, E., and Salinas, M. (1985). The mode of action of restrictocin and mitogillin on eukaryotic ribosomes. Inhibition of brain protein synthesis, cleavage and sequence of the ribosomal RNA fragment. *Eur. J. Biochem.*, **149**, 29–34.

Fernández-Puentes, C. and Carrasco, L. (1980). Viral infection permeabilizes mammalian cells to protein toxins. *Cell*, **20**, 769–75.

Gasset, M., Mancheño, J. M., Lacadena, J., Turnay, J., Olmo, N., Lizarbe, M. A., et al. (1994). α-Sarcin, a ribosome-inactivating protein that translocates across the membrane of phospholipid vesicles. *Curr. Topics Peptide Prot. Res* **1**, 99–104.

Goldin, A., Serpik, A., and Mantel, N. (1966). Experimental screening procedures and clinical predictability value. *Cancer Chem. Rep.*, **50**, 173–218.

Lacadena, J., Martínez del Pozo, A., Barbero, J. L., Mancheño, J. M., Gasset, M., Oñaderra, M., *et al.* (1994). Overproduction and purification of biologically active native fungal α-sarcin in *Escherichia coli*. *Gene,* **142**, 147–51.

Lamy, B. and Davies, J. (1991). Isolation and nucleotide sequence of the *Aspergillus restrictus* gene coding for the ribonucleolytic toxin restrictocin and its expression in *Aspergillus nidulans*: the leader sequence protects producing strains from suicide. *Nucleic Acid Res.*, **19**, 1001–6.

Lamy B., Moutaouakil, M., Latge, J.-P., and Davies, J. (1991). Secretion of a potential virulence factor, a fungal ribonucleotoxin, during human aspergillosis infections. *Mol. Microbiol.*, **5**, 1811–15.

Mancheño, J. M., Gasset, M., Lacadena, J., Martínez del Pozo, A., Oñaderra, M., and Gavilanes, J. G. (1995). Predictive study of the conformation of the cytotoxic protein α-sarcin. A structural model to explain α-sarcin–membrane interaction. *J. Theor. Biol.*, **172**, 259–67.

Martínez del Pozo, A., Gasset, M., Oñaderra M., and Gavilanes, J. G. (1988). Conformational study of the antitumour protein α-sarcin. *Biochim. Biophys. Acta*, **953**, 280–8.

Moser, M., Crameri, R., Menz, G., Schneider, T., Dudler, T., Virchow, C., *et al.* (1992). Cloning and expression of recombinant *Aspergillus fumigatus* allergen I/a (rAsp f I/a) with IgE binding and type I skin test activity. *J. Immunol.*, **149**, 454–60.

Oka, T., Natori, Y., Tanaka, S., Tsuguri, K., and Endo, Y. (1990). Complete nucleotide sequence of cDNA for the cytotoxin α-sarcin. *Nucleic Acid Res.*, **18**, 1897.

Olmo, N., Turnay, J., Lizarbe, M. A., and Gavilanes, J. G. (1993). Cytotoxic effect of α-sarcin, a ribosome inactivating protein, in culture Rugli cells. *S.T.P. Pharma Sci.*, **3**, 93–6.

Olson, B. H. and Goerner, G. L. (1965). α-Sarcin, a new antitumor agent. I. Isolation, purification, chemical composition and the identity of a new amino acid. *Appl. Microbiol.*, **13**, 314–21.

Otero, M. J. and Carrasco, L. (1986). External ATP permeabilizes transformed cells to macromolecules. *Biochem. Biophys. Res. Commun.*, **134**, 453–60.

Otero, M. J. and Carrasco, L. (1988). Exogenous phospholipase C permeabilizes mammalian cells to proteins. *Exp. Cell Res.*, **177**, 154–61.

Roga, V., Hedeman, L. P., and Olson, B. H. (1971). Evaluation of mitogillin (NSC-629529) in the treatment of naturally occurring canine neoplasms. *Cancer Chemother. Rep. Part 1*, **55**, 101–13.

Turnay, J., Olmo, N., Jiménez, A., Lizarbe, M. A., and Gavilanes, J. G. (1993). Kinetic study of the cytotoxic effect of α-sarcin, a ribosome inactivating protein from *Aspergillus giganteus* on tumour cell lines: protein biosynthesis inhibition and cell binding. *Mol. Cell. Biochem.*, **122**, 39–47.

Wawrzynczak, E. J., Henry, R. V., Cumber, A. J., Parnell, G. D., Derbyshire, E. J., and Ulbrich, N. (1991). Biochemical, cytotoxic and pharmacokinetic properties of an immunotoxin composed of a mouse monoclonal antibody Fib75 and the ribosome-inactivating protein α-sarcin from *Aspergillus giganteus*. *Eur. J. Biochem.*, **196**, 203–9.

Wool, I. G. (1984). The mechanism of action of the cytotoxic nuclease α-sarcin and its use to analyze ribosome structure. *Trends Biochem. Sci.*, **9**, 14–17.

Wool, I. G., Glück, A., and Endo, Y. (1992). Ribotoxin recognition of ribosomal RNA and a proposal for the mechanism of translocation. *Trends Biochem. Sci.*, **17**, 266–9.

■ *María Gasset:*
Instituto de Química-Física Rocasolano,
CSIC,
Serrano 119, 28006 Madrid,
Spain

4

Cytoskeleton-affecting toxins

Introduction

■ Cytoskeleton

Many cellular functions depend on the cytoskeleton. The cytoskeleton controls the shape and the spatial organization of the cell. It participates in all kinds of cellular movement and transport and is involved in processes like endo- and exocytosis, vesicle transport, cell–cell contact, and mitosis. The cytoskeleton consists of a fibre network that is formed of three major filament systems which are highly dynamic structures: the microfilaments, the microtubules, and the intermediate filaments. Rapid structural changes of these cytoskeletal proteins are based on their ability to polymerize and depolymerize. Much of what we know about functions of these proteins has been learned from the use of cytotoxins which have been applied as cell biological tools. The group of cytoskeleton-affecting toxins comprises agents which act directly on elements of the cytoskeleton but also toxins that affect regulatory components (e.g. low molecular mass GTP-binding proteins) which control the organization of the cytoskeleton. The toxins can be subdivided into protein toxins with enzymatic activity, protein toxins without apparent enzymic activity, and small non-proteinaceous compounds which bind to the cytoskeleton proteins. The bacterial protein toxins appear to act predominantly on the actin cytoskeleton.

■ Microfilaments and toxins

Microfilaments have a diameter of 7–9 nm and are built of polymerized actin. Actin is a 42 kDa protein of 375 amino acids which consists of four domains (I, II, III, IV), binds Mg^{2+} and ADP or ATP, and possesses ATPase activity (Kabsch and Vandekerckhove 1992). At least six actin isoforms are known in mammals, which are more than 95 per cent identical. For the cytoskeleton nonmuscle β/γ-actin appears to be most important.

Actin filaments are organized in linear bundles, two-dimensional networks, and three-dimensional gels. Actin is spread all over the cells but the highest concentration is submembraneously located forming the cell cortex. In cultured cells (e.g. fibroblasts), actin filaments are basically involved in the formation of finger-like spikes (filopodia) and broad membrane sheets (lamellipodia). These processes start from the leading edge and attach to the matrix by adhesion plaques. Moreover, independent from the matrix, actin filaments form ruffles. At these different sites, microfilaments are stabilized by a large family of actin-binding proteins. For example, at the adhesion plaques, actin is colocalized with vinculin, talin, tensin, and α-actinin (Burridge et al. 1988; Bretscher 1991).

Actin filaments are polar structures with a plus (fast-polymerizing) and a minus (slow-polymerizing) end. Deduced from the microscopic features of filaments after decoration with S1 subunits of myosin, actin filament ends are also known as barbed and pointed ends. At the plus end, actin tends to polymerize and at the pointed, it tends to depolymerize. Polymerization is largely affected by actin binding proteins like profilin and β-4-thymosine, which induce and inhibit ADP/ATP exchange, respectively, thereby stimulating or blocking polymerization (Nachmias 1993; Theriot and Mitchison 1993; Pollard et al. 1994). Moreover, various actin severing proteins have been identified which cut F-actin and regulate the length of actin filaments in a Ca^{2+} dependent or independent manner. Some of these proteins (e.g. gelsolin), keep bound to actin ends, thereby capping the filaments and blocking actin polymerization (Bershadsky and Vasiliev 1988; Vandekerckhove 1990).

Control of the actin cytoskeleton involves low molecular mass GTP-binding proteins of the Rho subtype family (Hall 1994; Machesky and Hall 1996; Narumiya 1996). Members of this protein family are RhoA,B,C, Rac1,2, Cdc42, RhoG, and TC10. Cdc42 appears to be involved in filopodia formation, Rac in membrane ruffling, Rho in formation of adhesion plaque and stress fibres. A regulatory cascade by these proteins has been suggested (Nobes and Hall 1995). The proteins are regulated by an inherent GTPase cycle (Bourne et al. 1990, 1991). In the active GTP-bound form the proteins interact with their effectors (e.g. protein kinases). The activated state is terminated by hydrolysis of GTP. Various regulatory proteins control the functions of Rho subtype proteins by blocking (GDI, guanine nucleotide dissociation inhibitor) or stimulating (GDS, guanine nucleotide dissociation stimulator) the nucleotide exchange or by activating the GTPase activity (GAP, GTPase activating proteins).

Actin is directly modified by *Clostridium botulinum* C2 toxin and the related iota-like toxins (*C. perfringens* iota toxin), which ADP-ribosylate the protein at Arg-177 thereby inhibiting actin polymerization and causing depolymerization of microfilaments (Aktories and Wegner 1992). These binary toxins are constructed according the A,B model and consist of a biologically active component (A) and a binding component (B), which is important for the binding at the cell surface and the translocation into the cell. In contrast to other A,B-toxins, the components of these binary toxins are separate proteins.

Other bacterial factors appear to act indirectly on the actin cytoskeleton. The so-called large clostridial cytotoxins (*C. difficile* toxins, *C. sordellii* toxins, and the α-toxin from *C. novyi*) and the C3-toxins modify small GTP-binding proteins (Rho proteins) which are involved in the regulation of the actin cytoskeleton by glyco-sylation and ADP-ribosylation, respectively (Aktories *et al.* 1992; Just *et al.* 1995). The cytotoxic necrotizing toxins (CNF1 and CNF2) from *E. coli* induce actin polymerization apparently by modification of Rho proteins (Oswald *et al.* 1994). ActA, which is produced by the *Listeria mono-cytogenes* is a bacterial cell surface protein which induces F-actin formation and enables the bacteria to move intra-cellularly (Domann *et al.* 1992). The zonula occludens toxin (ZOT) from *Vibrio cholerae* is an enterotoxin which acts on epithelial tight junctions to increase permeability (Baudry *et al.* 1992).

The nonproteinaceous toxins, which act on the actin cytoskeleton are cytochalasins and phalloidin (Cooper 1991). Cytochalasins are a family of fungal products which reversibly bind – somewhat similarly to capping proteins – to the barbed ends of actin filaments to block fast-actin polymerization. Like capping proteins they can nucleate polymerization. Therefore, in general, a com-plete depolymerization of actin is not achieved with cytochalasins. Cytochalasin B additionally inhibits sugar transport. Phallotoxins are cyclic oligopeptides which are isolated from *Amanita phalloides* (Wieland 1977). They bind to actin and decrease the critical concentration for actin polymerization thereby inducing and stabilizing microfilaments. Phallotoxins are not able to enter most cells but are taken-up by hepatocytes. Tagged with a fluorescence marker, they are widely used to stain F-actin by fluorescence microscopy (Faulstich *et al.* 1988). Jasplakinolide, which is a macrocyclic peptide produced by a marine sponge, decreases the critical concentration for actin polymerization in a manner similar to phallo-toxins (Bubb *et al.* 1994). This agent is able to enter cells. Swinholde A is another cytotoxic agent from a marine sponge. This cytotoxin binds to actin dimers and severs F-actin (Bubb *et al.* 1995).

Table 1 Microfilament-affecting toxins
(a) Protein toxins

Toxin	Origin	Structure	Mechanism	Effect
C2 toxin Iota toxin Spiroforme toxin	*C. botulinum* type C,D *C. perfringens* *C. spiroforme*	Binary protein toxins	ADP-ribosylation of actin	Inhibition of actin polymerization, depolymerization of F-actin
Toxin A Toxin B Lethal (LT) toxin	*C. difficile* *C. difficile* *C. sordellii*	Single chain protein toxins	Glucosylation of Rho, Rac, Cdc42 (LT also Ras, Rap, Ral, but not Rho)	Depolymerization of F-actin in cells
α-Toxin	*C. novyi*		*N*-acetylglucose-aminylation of Rho, Rac ,Cdc42	Depolymerization of F-actin in cells
C3 toxin Limosum toxin EDIN Cereus toxin	*C. botulinum* type C,D *C. limosum* *S. aureus* *B. cereus*	Single chain toxins, no binding component	ADP-ribosylation of Rho	Depolymerization of F-actin in cells
Cytotoxic necrotizing factor 1 and 2 (CNF1,2)	*E. coli*	Single chain protein toxins	Modification of Rho proteins (?)	Actin polymerization in cells
Zonula occludens toxin (ZOT)	*V. cholerae*	Enterotoxin	Unknown	Increase in permeability of tight junctions
ActA	*Listeria monocytogenes*	Bacterial cell surface protein	Unknown	Polymerization of actin in cells

(b) Nonproteinaceous toxins

Toxin	Origin	Structure	Mechanism	Effect
Cytochalasins (A,B,C,D,E,H)	*Aspergillus clavatus* and various other fungi	Macrolide-like isoindol derivatives	Actin binding, capping function	Inhibition of actin polymerization
Phallotoxins Virotoxins	*Amanita phalloides*	Bicyclic heptapeptides	actin binding, de-crease in the critical concentration of actin	Polymerization of actin
Jasplakinolide	*Jaspis johnstoni*	Cyclic peptide	Actin binding, decrease in the critical concentration of actin	Polymerization of actin
Swinholde A	*Theonella swinhoi*	Macrolide	Severing of F-actin	Depolymerization of F-actin

Microtubules and toxins

Microtubules have a diameter of 25 nm and consist of tubulin α/β heterodimers, which are often organized in 13 linear protofilaments forming a cylindric structure (Bershadsky and Vasiliev 1988). Therefore microtubules are more rigid than actin filaments. In mammals several tubulin isoforms exist, they consist of ~450 amino acids showing ~40 per cent identity. Whereas α-tubulin binds nonexchangeable GTP, β-tubulin has bound exchangeable GTP which is hydrolysed during polymerization. Microtubules are polar structures and contain a plus (fast-growing) and a minus (slow-growing or depolymerizing) end. Microtubules form parallel and radial arrays which often start at the microtubule organization centre (MTOC, centrosome). Various microtubule associated proteins (MAPs) modify the dynamic properties of tubulin. Microtubules are the scaffold for kinesins and dyneins, which are motor proteins for the intracellular transport to the plus and minus end, respectively (Gelfand and Bershadsky 1991).

In contrast to actin, no protein toxins are known which specifically affect microtubules. However, microtubules are targets for various nonproteinaceous toxins. Colchicine an alkaloid from the plant *Colchicum autumnale*, the related colcemid, nocadazole, podophyllotoxin, and vinca alkaloids inhibit polymerization or induce depolymerization of microtubules (Bergen and Borisy 1983; Goddette and Frieden 1986; Jordan *et al.* 1986). In contrast, taxol, a complex diterpene, isolated from certain types of the yew tree (*Taxus brevifolia*) induces tubulin polymerization even in the absence of GTP and microtubule-associated proteins, which are usually necessary for microtubule formation (Horwitz 1992).

Intermediate filaments

Intermediate filaments have a diameter of 10 nm and are built by a large spectrum of different protein classes (acidic and basic cytokeratins, vimentin/desmin, neurofilaments, and lamins), which form α-helical structures which assemble to rope-like fibres. Intermediate filaments are most important for the mechanical support of the cell and plays major role in cell–cell and cell–matrix contact. No evidence exists that intermediate filaments participate in motile functions (Bershadsky and Vasiliev 1988; Fuchs and Weber 1994).

So far, no toxins were shown to inhibit the functions of intermediate filaments. However, secondary to microfilament disruption by bacterial toxin, intermediate filaments are redistributed (Wiegers *et al.* 1991). The intermediate filament vimentin was identified to be an *in vitro* substrate for ADP-ribosylation by *Pseudomonas* exoenzyme S, which also modifies several different small GTP-binding proteins (Coburn *et al.* 1989).

References

Aktories, K. and Wegner, A. (1992). Mechanisms of the cytopathic action of actin-ADP-ribosylating toxins. *Mol. Microbiol.*, **6**, 2905–8.

Aktories, K., Mohr, C., and Koch, G. (1992). Clostridium botulinum C3 ADP-ribosyltransferase. *Curr. Top. Microbiol. Immunol.*, **175**, 115–31.

Baudry, B., Fasano, A., Ketley, J., and Kaper, J. B. (1992). Cloning of a gene (*zot*) encoding a new toxin produced by *Vibrio cholerae*. *Infect. Immun.*, **60**, No.2, 428–34.

Bergen, L. G. and Borisy G. G. (1983). Tubulin–colchicine complex inhibits microtubule elongation at both plus and minus ends. *J. Biol. Chem.*, **258**, 4190–4.

Bershadsky, A. D. and Vasiliev, J. M. (1988). *Cytoskeleton,* Plenum Press, New York.

Bourne, H. R., Sanders, D. A., and McCormick, F. (1990). The GTPase superfamily: a conserved switch for diverse cell functions. *Nature,* **348**, 125–32.

Bourne, H. R., Sanders, D. A., and McCormick, F. (1991). The GTPase superfamily: conserved structure and molecular mechanism. *Nature,* **349**, 117–27.

Bretscher, A. (1991). Microfilament structure and function in the cortical cytoskeleton. *Annu. Rev. Cell Biol.*, **7**, 337–74.

Bubb, M. R., Senderowicz, A. M., Sausville, E. A., Duncan, K. L. K., and Korn, E. D. (1994). Jasplakinolide, a cytotoxic natural product, induces actin polymerization and competively inhibits binding of phalloidin to F-actin. *J. Biol. Chem.*, **269**, 14869–71.

Bubb, M. R., Spector, I., Bershadsky, A. D., and Korn, E. D. (1995). Swinholide A is a microfilament disrupting marine toxin that stabilizes actin dimers and severs actin filaments. *J. Biol. Chem.*, **270**, 3463–6.

Burridge, K., Fath, K., Kelly, T., Nuckolls, G., and Turner, C. (1988). Focal adhesions: Transmembrane junctions between the

Table 2 Microtubule affecting toxins

Toxin	Origin	Structure	Mechanism	Effect
Taxol	*Taxus brevifolia*	Complex diterpene derivative	Tubulin binding	Polymerization of tubulin
Colchicine Colcemid	*Colchicum autumnale*	Tropolone derivative	Tubulin binding	Depolymerization of tubulin
Vinca alkaloids (e.g.Vinblastin)	*Catharanthus roseus*	Monoterpene-indole alkaloid compl.	Tubulin binding	Depolymerization of tubulin
Podophyllotoxins	*Podophyllum peltatum*	Phenylpropane derivative	Tubulin binding	Depolymerization of tubulin
Nocodazol		Benzimidazole derivative	Tubulin binding	Depolymerization of tubulin

extracellular matrix and the cytoskeleton. *Annu. Rev. Cell Biol.*, **4**, 487–525.

Coburn, J., Dillon, S. T., Iglewski, B. H., and Gill, D. M. (1989). Exoenzyme S of *Pseudomonas aeruginosa* specifically ADP-ribosylates the intermediate filament protein vimentin. *Infect. Immun.*, **57**, 996–8.

Cooper, J. M. (1991). Effects of cytochalasin and phalloidin on actin. *J. Cell Biol.*, **105**, 1473–8.

Domann, E., Wehland, J., Rohde, M., Pistor, S., Hartl, M., Goebel, W., *et al.* (1992). A novel bacterial virulence gene in *Listeria monocytogenes* required for host cell microfilament interaction with homology to the proline-rich region of vinculin. *EMBO J.*, **11**, 1981–90.

Faulstich, H., Zobeley, S., Rinnerthaler, G., and Small, J. V. (1988). Fluorescent phalloidtoxins as probes for filamentous actin. *J. Muscle Res. Cell Motil.*, **9**, 370–83.

Fuchs, E. and Weber, K. (1994). Intermediate filaments: Structure, dynamics, function, and disease. *Annu. Rev. Biochem.*, **63**, 345–82.

Gelfand, V. I. and Bershadsky, A. D. (1991). Microtubule dynamics: mechanism, regulation, and function. *Annu. Rev. Cell Biol.*, **7**, 93–116.

Goddette, D. W. and Frieden, C. (1986). The kinetics of cytochalasin D binding to monomeric actin. *J. Biol. Chem.*, **261**, 15970–3.

Hall, A. (1994). Small GTP-binding proteins and the regulation of the actin cytoskeleton. *Annu. Rev. Cell Biol.*, **10**, 31–54.

Horwitz, S. B. (1992). Mechanism of action of taxol. *Trends Pharmacol. Sci.*, **13**, 134–6.

Jordan, M. A., Margolis, R. L., Himes, R. H., and Wilson, L. (1986). Identification of a distinct class of vinblastine binding sites on microtubules. *J. Mol. Biol.*, **187**, 61–73.

Just, I., Selzer, J., Wilm, M., Von Eichel-Streiber, C., Mann, M.. and Aktories, K. (1995). Glucosylation of Rho proteins by *Clostridium difficile* toxin B. *Nature*, **375**, 500–3.

Kabsch, W. and Vandekerckhove, J. (1992). Structure and function of actin. *Annu. Rev. Biophys. Biophys. Chem.*, **21**, 49–76.

Machesky, L. M. and Hall, A. (1996). Rho: a connection between membrane receptor signalling and the cytoskeleton. *Trends Cell Biol.*, **6**. 304–10.

Nachmias, V. T. (1993). Small actin-binding proteins: The b-thymosin family. *Curr. Opin. Cell Biol.*, **5**, 56–62.

Narumiya, S. (1996). The small GTPase Rho: cellular functions and signal transduction. *J. Biochem.*, **120**, 215–28.

Nobes, C. D. and Hall, A. (1995). Rho, Rac, and Cdc42 GTPases regulate the assembly of multimolecular focal complexes associated with actin stress fibers, lamellipodia, and filopodia. *Cell*, **81**, 53–62.

Oswald, E., Sugai, M., Labigne, A., Wu, H. C., Fiorentini, C., Boquet, P., *et al.* (1994). Cytotoxic necrotizing factor type 2 produced by virulent *Escherichia coli* modifies the small GTP-binding proteins Rho involved in assembly of actin stress fibers. *Proc. Natl. Acad. Sci. USA*, **91**, 3814–18.

Pollard, T. D., Almo, S., Quirk, S., Vinson, V., and Lattman, E. E. (1994). Structure of actin binding proteins: Insights about function at atomic resolution. *Annu. Rev. Cell Biol.*, **10**, 207–49.

Theriot, J. A. and Mitchison, T. J. (1993). The three faces of profilin. *Cell*, **75**, 835–8.

Vandekerckhove, J. (1990). Actin-binding proteins. *Curr. Opin. Cell Biol.*, **2**, 41–50.

Wiegers, W., Just, I., Müller, H., Hellwig, A., Traub, P., and Aktories, K. (1991). Alteration of the cytoskeleton of mammalian cells cultured *in vitro* by *Clostridium botulinum* C2 toxin and C3 ADP-ribosyltransferase. *Eur. J. Cell Biol.*, **54**, 237–45.

Wieland, T. (1977). Modification of actins by phalloidtoxins. *Naturwissenschaften*, **64**, 303–9.

■ *Klaus Aktories:*
Institut für Pharmakologie und Toxikologie,
Albert Ludwigs Universität Freiburg,
Pharmakologisches Institut,
Herman Herderstrasse 5,
D-79104 Freiburg,
Germany

C2 toxin (*Clostridium botulinum* type C and D)

Clostridium botulinum *C2 toxin is a member of a family of binary cytotoxins that ADP-ribosylate monomeric G actin in arginine-177. Actin modification results in inhibition of actin polymerization and destruction of the actin cytoskeleton.*

C2 toxin is binary in structure and consists of a binding component (C2II, 100 kDa) and an enzyme component (C2I, 50 kDa) (Aktories *et al.* 1992). Both components are separated proteins. The gene of the enzyme component has been cloned recently (Fujii *et al.* 1996; EMBL Data Library D63903). C2II has to be activated by trypsin treatment to release an 75 kDa active fragment. C2II binds to a membrane receptor (not identified), which is present on all cell types studied so far. Binding of C2II induces a binding site for C2I. The toxin is taken-up by receptor-mediated endocytosis (Simpson 1989) (see Fig. 1). In artificial membranes, C2II induces cation selective and voltage-dependent channels (Schmid *et al.* 1994). C2I is an ADP-ribosyltransferase ($K_{m,NAD}$ = 4 μM) (Aktories *et al.* 1986) that modifies monomeric G-actin but not polymerized F-actin. Protein substrates are non-muscle β/γ actin and smooth muscle γ-actin, but not α-actin isoforms. Modification occurs specifically in arginine-177 (Vandekerckhove *et al.* 1988), which is located in an actin–actin contact site. ADP-ribosylation inhibits actin polymerization (Aktories *et al.* 1986) and blocks actin ATPase activity. ADP-ribosylated actin binds to the 'fast-polymerizing' (barbed) ends of actin filaments in a capping protein-like manner to block polymerization of nonmodified actin (Wegner and Aktories 1988). ADP-ribosylated actin does not interact with the 'slow-polymerizing' (pointed) ends of actin filaments allowing depolymerization of actin (Fig. 1). Moreover, ADP-ribosylation of actin inhibits the nucleation activity of the gelsolin-actin complex (Wille *et al.* 1992).

Related to C2 toxin are *Clostridium perfringens* iota toxin, *Clostridium spiroforme* toxin, and an ADP-ribosyltransferase produced by *Clostridium difficile* (Considine and Simpson 1991). All these toxins are binary in structure. The gene for iota toxin has been sequenced (Perelle *et al.* 1993) [EMBL Data Library X73562]. The membrane receptors for iota toxin and C2 toxin are different. The binding components of *Clostridium perfringens* iota toxin and of *Clostridium spiroforme* toxin are interchangeable but not with that of C2II. Iota toxin modifies all actin isoforms including α-isoforms in Arg-177.

■ Purification and sources

The purification procedure reported for C2 toxin from *Clostridium botulinum* type C strain 92-13 (produces no neurotoxin) is based on ammonium sulfate precipitation, DEAE-Sephadex, CM-Sephadex, and gel filtration (Ohishi *et al.* 1980). We isolate C2II by CM-sephadex and gel filtration. C2I is purified from the ammonium sulfate precipitate by isoelectric focusing (pI ~4.5) and gel filtration.

■ Toxicity

The single components of C2 toxin are almost nontoxic. The LD_{50} of the activated holotoxin (C2I plus C2II, ratio 1:2) is about 5 and 50 ng per mouse for i.v. and i.p. administration, respectively (Ohishi *et al.* 1980). In rats C2 toxin induces hypotension, hemorrhaging and fluid accumulation into the lungs (Simpson 1982). Major toxic effects may be explained by massive increase in vascular permeability. C2 toxin is not a clostridial neurotoxin and plays no role in botulism. In intact cells, C2 toxin induces rounding-up of cells, destruction of the microfilament network, and complete depolymerization of actin filaments (Reuner *et al.* 1987). Effects on actin cytoskeleton occur with a latency (toxin up-take) of about 30 min.

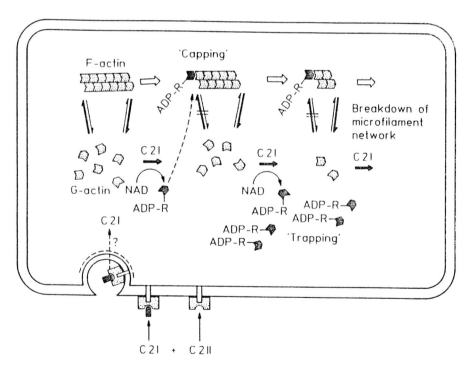

Figure 1. Model of the cytopathic effects of C2 toxin. The activated binding component (C2II) binds to the cell membrane receptor and induces the binding site for the enzyme component (C2I). After receptor-mediated endocytosis and translocation of C2I into the cytoplasma, actin is ADP-ribosylated. ADP-ribosylation inhibits actin polymerization. Furthermore, ADP-ribosylated actin acts like a capping protein to inhibit polymerization at the barbed ends of filaments. Depolymerization of F-actin occurs at the pointed ends of microfilaments (from Aktories *et al.* 1992).

Use in cell biology

C2 toxin is the most effective agent to depolymerize F-actin in intact cells (Aktories and Wegner 1992) and is applied to study the functional roles of the actin cytoskeleton (Aktories et al. 1992). In neutrophil leukocytes, C2 toxin inhibits migration and increases superoxide anion production and enzyme secretion induced by chemotactic agents (e.g. formyl-methionine-leucine-phenylalanine, FMLP) (Norgauer et al. 1988). FMLP-induced actin polymerization is inhibited and F-actin content is decreased by > 75 per cent. Endocytosis of the N-formyl peptide receptor is slowed down but still possible. C2 toxin augments ligand-induced phosphoinositide hydrolysis and increases diacylglycerol production several-fold and largely affects production of lipid mediators (Grimminger et al. 1991). The effects of C2 toxin were studied on catecholamine release in PC-12 cells (Matter et al. 1989), on steroid release in Y-1 cells (Considine et al. 1992), and on insulin secretion in pancreatic islets and HIT-T15 cells (Li et al. 1994). C2 toxin increases permeability and hydraulic conductivity in monolayers of endothelium cells (Suttorp et al. 1991). In isolated longitudinal muscle of guinea pig ileum, smooth muscle contraction is inhibited by C2 toxin (Mauss et al. 1989).

References

Aktories, K. and Wegner, A. (1992). Mechanisms of the cytopathic action of actin-ADP-ribosylating toxins. *Mol. Microbiol.*, **6**, 2905–8.

Aktories, K., Bärmann, M., Ohishi, I., Tsuyama, S., Jakobs, K. H., and Habermann, E. (1986). Botulinum C2 toxin ADP-ribosylates actin. *Nature*, **322**, 390–2.

Aktories, K., Wille, M., and Just, I. (1992). Clostridial actin-ADP-ribosylating toxins. *Curr. Top. Microbiol. Immunol.*, **175**, 97–113.

Considine, R. V. and Simpson, L. L. (1991). Cellular and molecular actions of binary toxins possessing ADP-ribosyltransferase activity. *Toxicon*, **29**, 913–36.

Considine, R. V., Simpson, L. L., and Sherwin, J. R. (1992). Botulinum C$_2$ toxin and steroid production in adrenal Y-1 cells: The role of microfilaments in the toxin-induced increase in steroid release. *J. Pharmacol. Exp. Ther.*, **260**, 859–64.

Fujii, N., Kubota, T., Shirakawa, S., Kimura, M., Ohishi, I., Moriishi, K., Isogai, E., and Isogai, H. (1996). Charaterization of component-I gene of botulinum C2 toxin and PCR detection of its gene in clostridial species. *Biochem. Biophys. Res. Commun.*, **220**, 353–9.

Grimminger, F., Sibelius, U., Aktories, K., Suttorp, N., and Seeger, W. (1991). Inhibition of cytoskeletal rearrangement by botulinum C2 toxin amplifies ligand-evoked lipid mediator generation in human neutrophils. *Mol. Pharmacol.*, **40**, 563–71.

Li, G., Rungger-Brändle, E., Just, I., Jonas, J.-C., Aktories, K., and Wollheim, C. B. (1994). Effect of disruption of actin filaments by *Clostridium botulinum* C2 toxin on insulin secretion in HIT-T15 cells and pancreatic islets. *Mol. Biol. Cell*, **5**, 1199–213.

Matter, K., Dreyer, F., and Aktories, K. (1989). Actin involvement in exocytosis from PC12 cells: studies on the influence of botulinum C2 toxin on stimulated noradrenaline release. *J. Neurochem.*, **52**, 370–6.

Mauss, S., Koch, G., Kreye, V. A. W., and Aktories, K. (1989). Inhibition of the contraction of the isolated longitudinal muscle of the guinea-pig ileum by botulinum C2 toxin: Evidence for a role of G/F-actin transition in smooth muscle contraction. *Naunyn-Schmiedebergs Arch. Pharmacol.*, **340**, 345–51.

Norgauer, J., Kownatzki, E., Seifert, R., and Aktories, K. (1988). Botulinum C2 toxin ADP-ribosylates actin and enhances O^{2-} production and secretion but inhibits migration of activated human neutrophils. *J. Clin. Invest.*, **82**, 1376–82.

Ohishi, I., Iwasaki, M., and Sakaguchi, G. (1980). Purification and characterization of two components of botulinum C2 toxin. *Infect. Immun.*, **30**, 668–73.

Perelle, S., Gibert, M., Boquet, P., and Popoff, M. R. (1993). Characterization of *Clostridium perfringens* iota-toxin genes and expression in *Escherichia coli*. *Infect. Immun.*, **61**, 5147–56.

Reuner, K. H., Presek, P., Boschek, C. B., and Aktories, K. (1987). Botulinum C2 toxin ADP-ribosylates actin and disorganizes the microfilament network in intact cells. *Eur. J. Cell Biol.*, **43**, 134–40.

Schmid, A., Benz, R., Just, I., and Aktories, K. (1994). Interaction of *Clostridium botulinum* C2 toxin with lipid bilayer membranes: formation of cation-selective channels and inhibition of channel function by chloroquine and peptides. *J. Biol. Chem.*, **269**, No.24, 16706–11.

Simpson, L. L. (1982). A comparison of the pharmacological properties of *Clostridium botulinum* type C1 and C2 toxins. *J. Pharmacol. Exp. Ther.*, **223**, 695–701.

Simpson, L. L. (1989). The binary toxin produced by *Clostridium botulinum* enters cells by receptor-mediated endocytosis to exert its pharmacologic effects. *J. Pharmacol. Exp. Ther.*, **251**, 1223–8.

Suttorp, N., Polley, M., Seybold, J., Schnittler, H., Seeger, W., Grimminger, F., et al. (1991). Adenosine diphosphate-ribosylation of G-actin by botulinum C2 toxin increases endothelial permeability in vitro. *J. Clin. Invest.*, **87**, 1575–84.

Vandekerckhove, J., Schering, B., Bärmann, M., and Aktories, K. (1988). Botulinum C2 toxin ADP-ribosylates cytoplasmic b/v-actin in arginine 177. *J. Biol. Chem.*, **263**, 696–700.

Wegner, A. and Aktories, K. (1988). ADP-ribosylated actin caps the barbed ends of actin filaments. *J. Biol. Chem.*, **263**, 13739–42.

Wille, M., Just, I., Wegner, A., and Aktories, K. (1992). ADP-ribosylation of the gelsolin–actin complex by clostridial toxins. *J. Biol. Chem.*, **267**, 50–5.

■ *Klaus Aktories:*
Institut für Pharmakologie und Toxikologie,
Albert Ludwigs Universität Freiburg,
Pharmakologisches Institut,
Hermann Herderstrasse 5,
D-79104 Freiburg,
Germany

Cytotoxic necrotizing factors (*Escherichia coli*)

Cytotoxic necrotizing factors (CNFs) are single-chain proteins of about 110 kDa elaborated by a number of pathogenic E. coli *strains. CNFs induce ruffling and stress fibre formation into cultured cells probably by activating the p21 Rho G-protein. Stimulation of actin assembly leads to the block of cytokinesis, resulting in multinucleation.*

Certain *Escherichia coli* strains belonging to the normal microflora present in the gastrointestinal tract, have been often associated with gastroenteritis, urogenital infections, septicemia, and bacteremia in both humans and animals. Among these pathogenic *E. coli,* a number of strains have been described as producing new putative virulence factors named cytotoxic necrotizing factors type 1 and 2 (CNF1 and CNF2). The first type of CNF was described as a cell-associated product of *E. coli* strains isolated from children with diarrhea causing necrosis of rabbit skin and multinucleation in cultured cells (Caprioli *et al.* 1983). Subsequently, a second type of CNF was detected in extracts of *E. coli* strains isolated from calves with enteritis (De Rycke *et al.* 1987). Purified CNF1 (Caprioli *et al.* 1984) and CNF2 (Oswald and De Rycke 1990) are immunologically related and similar in molecular weights (113.7 and 110 kDa, respectively). However, the toxins are distinguishable by:

(1) the morphology of multinucleated cells induced by each toxin in HeLa cell cytotoxic assays;

(2) their different responses in cross-neutralization assays; and

(3) the specific necrotic activity of CNF2 in the mouse footpads (De Rycke *et al.* 1990). Experimental infections of neonatal calves and pigs (Wray *et al.* 1993) have shown that orally inoculated CNF1- or CNF2-producing *E. coli* cause septicemia, enteritis, and histological changes characteristic of toxemic effects in the brain, heart, liver, and kidney. These lesions were similar to those observed after the intravenous inoculation of purified CNF1 in lambs (De Rycke and Plassiart 1990).

■ Purification

CNF1 and CNF2 are accumulated in the cytosol of producing bacteria during the exponential growth phase. CNFs-producing strains grown in LB medium, were harvested at the end of the exponential growth phase, centrifuged, and resuspended in 50 mM sodium phosphate buffer (pH 7.4). Bacteria were then disrupted by French press, centrifuged at high speed, and resuspended; the supernatant was precipitated with 50 per cent ammonium sulfate. The purification scheme for both toxins involves five main steps, as reported in Donelli *et al.* (1994). Briefly, after DEAE ion exchange, the fractions with toxin activity were pooled and purified by two consecutive gel filtrations, first on ACA54 and then on ACA44 columns. After a passage through a C6-Sepharose column, the fractions containing CNFs activity were pooled and purified by a monoQ FPLC column. CNFs purity was assessed by SDS-PAGE.

■ Molecular genetics

CNF1 is chromosomally encoded (Falbo *et al.* 1992), whereas the determinant for CNF2 is located on a large transmissible F-like plasmid called Vir (Oswald and De Rycke 1990). CNF1-producing strains isolated from diarrheal diseases and extraintestinal infections in humans, calves, pigs, cats, and dogs (Caprioli *et al.* 1983, 1987; De Rycke *et al.* 1987; Prada *et al.* 1987; Cherifi *et al.* 1990; Blanco *et al.* 1993) produce an alpha-hemolysin and induce mannose resistant hemagglutination. DNA hybridization with probes derived from a Pap adhesin operon and isolation of cosmids with a DNA fragment containing both *cnf1*, the gene encoding CNF1, and the alpha-hemolysin operon (Falbo *et al.* 1992) support the hypothesis that the genes of these three virulence factors (alpha-hemolysin, Pap-like adhesin, and CNF1) are closely associated on the chromosome. Most of the CNF2-producing strains are nonhemolytic, do not possess a Pap-like adhesin but carry virulence determinants for colonization of the calf and lamb intestine (Falbo *et al.* 1993). In fact, CNF2-producing strains are only isolated from the blood of diarrheal stools of domestic polygastric animals (De Rycke *et al.* 1987; Oswald *et al.* 1991; Blanco *et al.* 1993).

Recently, the genes encoding CNFl (*cnf1*) and CNF2 (*cnf2*) have been cloned (Falbo *et al.* 1993; Oswald *et al.* 1994) and the sequences of both genes are available (cnf1: accession no. X70670; cnf2: accession no. U01097). Both toxins are encoded by a single structural gene with a low GC content (35 per cent). No classical signal sequence was found in the N-terminal 50 residues of these toxins. When the deduced amino acid sequences of the two toxins were compared, 85 per cent identical and

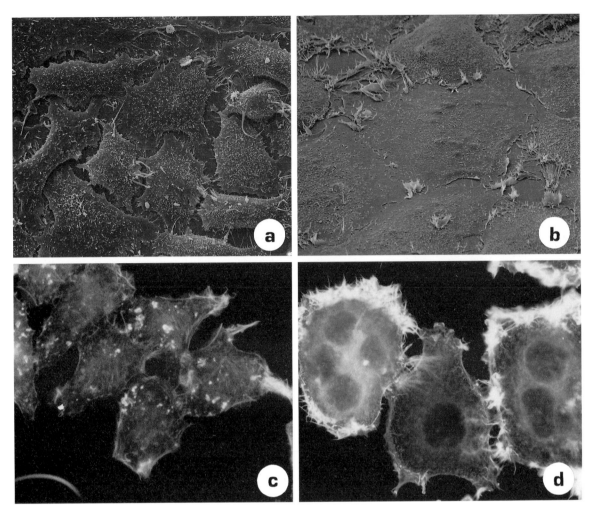

Figure 1. Scanning electron (a, b) and fluorescence (c, d) micrographs of HEp-2 cells treated with CNF1 (10^{-10}) M for 48 h. (Untreated (a, c) and CNF1-treated (b, d) cells.) Cells exposed to the toxin showed (b) the enlargement and flattening of the cell body, and (d) multinucleation accompanied by an intensive membrane ruffling positive for F-actin.

99 per cent conserved residues over 1014 amino acids were found. Nucleotide and protein data base searches showed significant homology between CNFs and the dermonecrotic toxin of *Pasteurella multocida* (Falbo *et al.* 1993; Oswald *et al.* 1994).

■ Use in cell biology

In cultured mammalian cells both CNFs act as classical bacterial toxins which have to be internalized by endocytosis in order to exert their effects (Falzano *et al.* 1993; Oswald *et al.* 1994). The two toxins are able to provoke a time- and dose-dependent increase in F-actin structures, inducing an intense and generalized ruffling activity and a remarkable increase in stress fibres (Fiorentini *et al.*

1988; Falzano *et al.* 1993; Oswald *et al.* 1994). Reorganization of the F-actin cytoskeleton by CNFs results in the inability of the cells to achieve correctly their cytokinesis and thus to divide normally, giving rise to extremely flat and large multinucleated cells.

Actin assembly/disassembly has been reported to be controlled by Rho proteins, a subgroup of small GTP-binding molecules belonging to the p21 Ras superfamily. Interestingly, CNFs are able to modify Rho by causing a shift of this protein towards a higher molecular weight (Oswald *et al.* 1994). A number of bacterial toxins, such as *Clostridium difficile* toxin B or *Clostridium botulinum* exoenzyme C3, are known to disrupt the actin cytoskeleton by inactivating the Rho protein (Donelli and Fiorentini 1994). However, in contrast to these toxins, CNF1 probably induces a covalent modification of Rho able to maintain this small GTP-binding protein in an activated form, thereby promoting F-actin assembly in the cell (Fiorentini *et al.* 1995).

■ References

Blanco, J., Blanco, M., Wong, I., and Blanco, J. E. (1993). Haemolytic *Escherichia coli* strains isolated from stools of healthy cats produce cytotoxic necrotizing factor type 1 (CNF1). *Vet. Microbiol.*, **38**, 157–65.

Caprioli, A., Falbo, V., Roda, L. G., Ruggeri, F. M., and Zona, C. (1983). Partial purification and characterization of an *Escherichia coli* toxic factor that induces morphological cell alterations. *Infect. Immun.*, **39**, 1300–6.

Caprioli, A., Donelli, G., Falbo, V., Possenti, R., Roda, L. G., Roscetti, G., *et al.* (1984). A cell division-active protein from *E. coli*. *Biochem. Biophys. Res. Commun.*, **118**, 587–93.

Caprioli, A., Falbo, V., Ruggeri, F. M., Baldassarri, L., Bisicchia, R., Ippolito, G., *et al.* (1987). Cytotoxic necrotizing factor production by hemolytic strains of *Escherichia coli* causing extraintestine infections. *J. Clin. Microbiol.*, **25**, 146–9.

Cherifi, A. M., Contrepois, M., Picard, B., Goullet, P., De Rycke J., Fairbrother, J., *et al.* (1990). Factor and markers of virulence in *Escherichia coli* from human septicemia. *FEMS Microbiol. Lett.*, **70**, 279–84.

De Rycke J. and Plassiart, G. (1990). Toxic effects for lambs of cytotoxic necrotizing factor from *Escherichia coli*. *Res. Vet. Sci.*, **49**, 349–54.

De Rycke, J., Guillot, J. F., and Boivin, R. (1987). Cytotoxins in non-enterotoxigenic strains of *Escherichia coli* isolated from feces of diarrheic calves. *Vet. Microbiol.*, **15**, 137–57.

De Rycke, J., Gonzàlez, E. A., Blanco, J., Oswald, E., Blanco, M., and Boivin, R. (1990). Evidence of two types of cytotoxic necrotizing factors (CNF1 and CNF2) in human and animal clinical isolates of *Escherichia coli*. *J. Clin. Microbiol.*, **28**, 694–9.

Donelli, G., and Fiorentini, C. (1994). Bacterial protein toxins acting on the cell cytoskeleton. *Microbiologica*, **17**, 345–62.

Donelli, G., Fiorentini, C., Falzano, L., Pouchelet, M., Oswald, E., and Boquet, P. (1994). Effects induced by the cytotoxic necrotizing factor 1 (CNF1) from pathogenic *E. coli* on cultured epithelial cells. *Zbl. Bakt. Suppl.*, **24**, 60–71.

Falbo, V., Famiglietti, M., and Caprioli, A. (1992). Gene block encoding production of cytotoxic necrotizing factor 1 and hemolysin in *Escherichia coli* isolates from extraintestinal infections. *Infect. Immun.*, **60**, 2182–7.

Falbo, V., Pace, T., Picci, L., Pizzi, E., and Caprioli, A. (1993). Isolation and nucleotide sequence of the gene encoding cytotoxic necrotizing factor 1 of *Escherichia coli*. *Infect. Immun.*, **61**, 4909–14.

Falzano, L., Fiorentini, C., Donelli, G., Michel, E., Kocks, C., Cossart, P., *et al.* (1993). Induction of phagocytic behaviour in human epithelial cells by *E. coli* cytotoxic necrotizing factor type 1. *Mol. Microbiol.*, **9**, 1247–54.

Fiorentini, C., Arancia, G., Caprioli, A., Falbo, V., Ruggeri, F. M., and Donelli, G. (1988). Cytoskeletal changes induced in HEp-2 cells by the cytotoxic necrotizing factor of *Escherichia coli*. *Toxicon*, **26**, 1047–56.

Fiorentini, C., Donelli, G., Matarrese, P., Fabbri, A., Paradisi, S., and Boquet, P. (1995). *Escherichia coli* cytotoxic necrotizing factor 1 (CNF1): evidence for induction of actin assembly by constitutive activation of the p21 Rho GTPase. *Infect. Immun*, **63**, 3936–44.

Oswald, E. and De Rycke, J. (1990). A single protein of 110 kDa is associated with the multinucleating and necrotizing activity coded by the Vir plasmid of *Escherichia coli*. *FEMS Microbiol. Lett.*, **68**, 279–84.

Oswald, E., De Rycke, J., Lintermans, P., Van Muylem, K., Mainil, J., Daube, G., *et al.* (1991). Virulence factors associated with cytotoxic necrotizing factor type two in bovine diarrheic and septicemic strains of *Escherichia coli*. *J. Clin. Microbiol.*, **29**, 2522–7.

Oswald, E., Motoyuki, S., Labigne, A., Wu, H. C., Fiorentini, C., Boquet, P., *et al.* (1994). Cytotoxic necrotizing factor type 2 produced by virulent *Escherichia coli* modifies the small GTP-binding proteins Rho involved in assembly of actin stress fibers. *Proc. Natl. Acad. Sci. USA*, **91**, 3814–18.

Prada, J., Baljer, G., De Rycke, J., Steinruck, H., Zimmerrnann, S., Stephan, R. *et al.* (1987). Characteristics of α-haemolyitc strains of *Escherichia coli* isolated from dogs with gastroenteritis. *Vet. Microbiol.*, **29**, 59–73.

Wray, C., Piercy, D. W. T., Carroll, P. J., and Cooley, W. A. (1993). Experimental infections of neonatal pigs with CNF toxin-producing strains of *Escherichia coli*. *Res. Vet. Sci.*, **54**, 290–8.

■ *Gianfranco Donelli and Carla Fiorentini:*
Department of Ultrastructures,
Istituto Superiore di Sanità,
Viale Regina Elena 299,
00161 Rome,
Italy

Enterotoxin A and cytotoxin B (*Clostridium difficile*)

Clostridium difficile enterotoxin A (TcdA, 308 kDa) and cytotoxin B (TcdB, 270 kDa) belong to the group of large clostridial cytotoxins (LCT). The toxins are secreted into the culture supernatant of the growing bacteria, specifically bind to eukaryotic cells, and are then taken up by receptor mediated endocytosis. Intracellularly they monoglucosylate small GTP-binding proteins, mainly of the Rho subfamily, at their effector domain. The GTPases are thus functionally inactivated, the result is a breakdown of the cellular actin stress fibres, a block of cytokinesis, but not a loss of vitality of the cells.

■ The human pathogen

The enterotoxin A (TcdA) and the cytotoxin B (TcdB) of *Clostridium difficile* are the two virulence factors responsible for the induction of antibiotical associated diarrhea (AAD) or pseudomembranous colitis (PMC) (Lyerley *et al.* 1988; Knoop *et al.* 1993). Diagnosis of the disease is done by identifying the toxins in stool specimens.

■ Purification and sources

The high molecular weight cytotoxins are secreted by the bacterium into its culture supernatants from which they may be isolated. One high producer (Sullivan *et al.* 1982), *C. difficile* VPI10463 (see Table 1), is used in many laboratories. The culture technique first introduced by Sterne and Wentzel (1950) is recommended for isolation. The central and final purification step is an ion-exchange chromatography most effectively done on a high resolution column like MonoQ (Eichel-Streiber *et al.* 1987). The other large clostridial cytotoxins (LCT, see Table 1) may

be purified by the same route. Starting from a 3 litre culture, good preparations yield a total of 6–18 mg TcdA (0.6–3 mg/ml) and 2–10 mg of TcdB (0.2–1.0 mg/ml). On SDS-PAGE gels the polypeptides migrate above the 200 kDa myosin standard protein. Major contaminating proteins are a 46 kDa glutamate dehydrogenase in TcdA (Lyerly *et al.* 1991) and a 150 kDa protein in TcdB (Pothoulakis *et al.* 1986). Special protocols have been described for TcdA purification using thyroglobulin affinity chromatography (Krivan and Wilkins 1987) and for TcdB purification using the addition of $CaCl_2$ as an additive during anion-exchange chromatography (Meador and Tweten 1988).

■ Antibodies

Polyclonal antisera against TcdA are almost exclusively directed against the C-terminal repetitive domain (Eichel-Streiber *et al.* 1989). Polyclonal antiserum against TcdB is difficult to raise. In many instances a polyclonal antiserum against the lethal toxin (TcsL) of *C. sordellii* proved to

Table 1 Properties of large clostridial cytotoxins

	Clostridium difficile			C. sordellii		C. novyi
Strain	VPI10463 (ATCC43255)		1470 (ATCC43598)	VPI9048 IP82		ATCC19402
Toxin (former abbreviation)	TcdA (ToxA)	TcdB (ToxB)	TcdB-1470 (TcdBF)	TcsH (HT)	TcsL (LT)	Tcnα (–)
Synonym	enterotoxin	cytotoxin	cytotoxin	hemorrhagic toxin	lethal toxin	alpha-toxin
M_r	308.000	270.000	269.000	300.000	250.000	258.000
P*I*	5.3	4.1	4.1	6.1	4.55	5.9
Cell culture cytotoxicity (ng/m)	10	1–50	10	15–500	1.6–16	0.1–10
Cytopathic effect	TcdB-like	TcdB	TcsL-like	TcsL-like	TcsL	TcdB-like
Mouse (lethal dose)	50–100 ng	50–100 ng	–	75–120 ng	3–5 ng	5–10 ng
Target GTPase	Rho, Rac Cdc42	Rho, Rac Cdc42	Rac, Rap	Rac, Ras, Rap, Ral	n. d.	Rho, Rac, Cdc42
mAbs	TTC8 PCG-4	2CV	2CV	TTC8 PCG-4	2CV	–
Gene	EMBL X51797	EMBL X53138	EMBL Z23277	–	EMBL X82638	EMBL Z48636

react effectively with TcdB (Eichel-Streiber et al. 1990). Although TcdA and TcdB are highly homologous to each other (Eichel-Streiber et al. 1992), properly diluted, neither polyclonal antiserum is crossreactive.

Two mAbs (TTC8 and PCG-4) against TcdA have been raised, they react with epitopes of the repeat structures (Lyerly et al. 1986; Sauerborn et al. 1994) and neutralize TcdA in vivo. TTC8 crossreacts with TcsH of C. sordellii. Only a few antibodies against TcdB exist (Muller et al. 1992a; Sauerborn et al. 1994). The best characterized is 2CV reacting with a sequence of the C-terminal repeat of TcdB (Sauerborn et al. 1994). The latter mAb does not neutralize the TcdB action. Again 2CV crossreacts with TcsL of C. sordellii.

■ Interaction with cells

The toxins are taken up by receptor mediated endocytosis (Eichel-Streiber et al. 1991). The repeat-domain of TcdA has been identified as the site of interaction with a carbohydrate (Galα1-3Galβ1-4GlcNAc) structure (Krivan and Wilkins 1987). A similar structure is present in thyroglobulin (Krivan et al. 1986) and also on a partially purified glycoprotein receptor molecule (Rolfe and Song 1993). Inactivation of TcdA with TTC8 is due to blocking the receptor–ligand interaction (Sauerborn and Eichel-Streiber 1995). No partner molecule for TcdB on the cell surface has yet been found. Due to their similar architecture an interaction of the TcdB repeat with another oligosaccharide is anticipated as the basis for its uptake.

During the passage of the toxins into the cell they are blocked by neutralization of the endosomal pH (Florin and Thelestam 1986; Henriques et al. 1986). Since both TcdA and TcdB are equally active when microinjected into the cell cytoplasm (Muller et al. 1992b), it was speculated that they do not need to be activated. The extreme amount of toxin needed for in vitro modification of the target proteins (Just et al. 1995), now indicates that such processing is indeed essential to deliver the full activity. Once inside the cell the toxins affect small GTP-binding proteins by an enzymatic mechanism described below (Just et al. 1995).

■ Genes and genetic variation

The toxin genes tcdA and tcdB together have been cloned and sequenced (Dove et al. 1990; Eichel-Streiber and Sauerborn 1990; Eichel-Streiber et al. 1990; Johnson et al. 1990). Together with three accessory genes (tcd C-E) they constitute the pathogenicity locus (PaLoc) of C. difficile (Braun et al. 1996). The single chain proteins are encoded by genes of 8130 and 7098 bp length (Eichel-Streiber 1993). Comparative analysis of a greater variety of strains proved that considerable genetic polymorphisms of the toxin genes exist (Rupnik et al. 1995). To date, for some isolates species-classification as C. difficile is disputed (Depitre et al. 1993). Since different LCTs modify a different spectrum of GTPases, dealing with 'C. difficile' toxins does not necessary mean that the toxins react with identical targets. If LCTs are used as tools in molecular cell biology this has to be taken into account.

■ Characterization of the protein

The molecular sizes and isoelectric points of TcdA and TcdB are 308 kDa, p/ 5.3 and 270 kDa, p/ 4.1, respectively (Eichel-Streiber 1993). The polypeptides are not modified by mercaptoethanol addition. Protease digestion does not lead to production of defined fragments but rather totally breaks the toxins into pieces (Eichel-Streiber, unpublished data).

The protein sequences derived from the tcdA and tcdB genes share 63 per cent homologous amino acid positions (Eichel-Streiber et al. 1992). Structurally they are both composed of two parts, the N-terminal nonrepetitive two thirds, and the C-terminal third which is highly repetitive (Eichel-Streiber et al. 1992; Eichel-Streiber 1993). The repeats have been designated CROPs (clostridial repetitive oligopeptides) (Eichel-Streiber et al. 1992; Eichel-Streiber 1993; Hofmann et al. 1995). This part of the toxin shares sequences with glucosyltransferases of Streptococci (Eichel-Streiber et al. 1992).

Sequencing gave raise to a three domain structure of the toxins, with a C-terminal repetitive area recognized as the ligand part by neutralizing mAbs (Sauerborn and Eichel-Streiber 1995), a central hydrophobic part encompassing several transmembranal segments (Hofmann et al. 1995), probably functioning as a translocation unit, and an N-terminal part which obviously encodes for the catalytic domain (Eichel-Streiber et al. 1995).

■ Large clostridial cytotoxins

C. difficile toxins are A–B type bacterial toxins belonging to the group of large clostridial cytotoxins (see Table 1). The other three members of LCTs are C. sordellii hemorrhagic (TcsH) and the lethal (TcsL) toxins and C. novyi α-toxin (Tcnα). The tcnα (Hofmann et al. 1995) and tcsL (Grenn et al. 1995) genes are sequenced, again CROPs were identified and both toxins are homologous to the C. difficile toxins TcdA and TcdB. The latter homology supports the notion that the toxins of C. sordellii and C. difficile are immunologically related (Popoff 1987; Martinez and Wilkins 1988). One C. difficile strain produces a cytotoxin B with a modified mode of action (Eichel-Streiber et al. 1995). The cytotoxic response of treated cells is that of TcsL of C. sordellii rather than that of TcdB of C. difficile (Eichel-Streiber et al. 1995). This again points to the possibility that more cytotoxins exist which should be placed into the LCT-group and that the spectrum of GTPases modified might be greater than defined today.

Different from the diphtheria-type of toxins equimolarly composed of one A and one B subunit (Choe et al. 1992), and cholera-type toxins being composed of one A and five B subunits (Sixma et al. 1991), C. difficile toxins have a single chain containing several ligand domains. Thus they form a novel third class of ABl toxins (Eichel-Streiber et al. 1996).

■ Catalytic activity

TcdA and TcdB are enzymes which monoglucosylate small GTP-binding proteins of the Ras superfamily (Just et al.

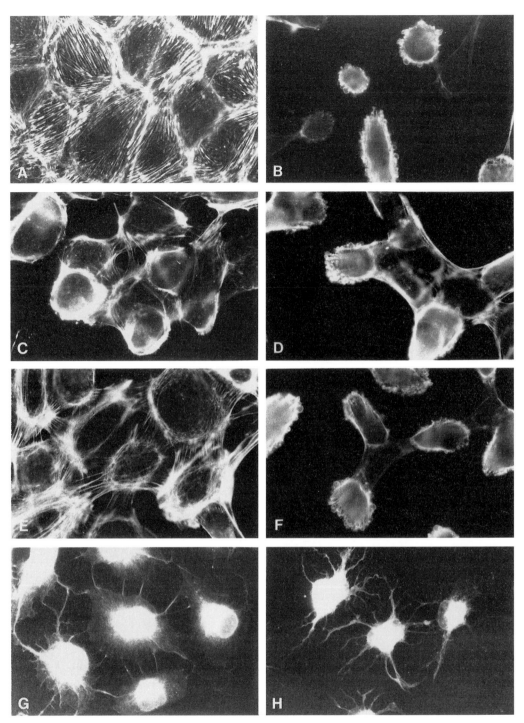

Figure 1. Distribution of actin filaments after intoxication with large clostridial cytotoxins. Endothelial cells from the pig pulmonary artery were taken into culture and incubated with different toxin concentrations for the times indicated. A, Untreated control; B–D, treatment with TcdB-1470: B, 1000 mg/ml 8 h; C, 200 ng/ml 4 h; D, 1000 ng/ml 4 h; E–F, treatment with *C. sordellii* TcsL: E, 200 ng/ml 4 h; F, 1000 ng/ml 4 h; G–H, treatment with TcdB of *C. difficile* VPI10463: G, 5 ng/ml 4 h; H, 200 ng/ml 4 h. Cells were stained with phalloidin-fluorescein to visualize F-actin.

1994, 1995). Three target proteins have been identified, Rho, Rac, and CDC42, all belonging to the Rho subfamily of GTPases. For Rho the site of modification is Thr-37 (Just et al. 1995). TcdB-glucosylated-Rho is no longer a substrate for the ADP-ribosyltransferase C3 of *C. botulinum* (Just et al. 1994). Transient overexpression of Rho mediates resistance of cells against *C. difficile* toxin A and B but not against *C. sordellii* lethal toxin (Giry et al. 1995).

The cofactor in the reaction is an activated glucose moiety (UDP-glucose) (Just et al. 1995). UDP-14C-glucose is solved in ethanol which has to be evaporated before the reaction is started. For Tcnα UDP-GlcNac is the cofactor (Selzer et al. 1995). The source of target GTPases is recombinant material or the pool of GTPases present in highly concentrated cell lysates. The reaction buffer is adapted to the intracellular medium. First GDP-loading of the GTPases is performed, thereafter toxin and cofactor are added and the labelling reaction is carried out. The reaction mixture is subjected to SDS-PAGE electrophoresis and radioactivity is measured in a phosphoimager.

The target proteins are GTPases of the Ras-superfamily of small GTP-binding proteins. The toxins analysed so far modify GTPases of the Rho- and Ras-subfamilies. Known targets of the individual toxins are listed in Table 1 (Eichel-Streiber et al. 1996; Hofmann et al. 1996; Just et al. 1995; Popoff et al. 1996; Selzer et al. 1996).

∎ *In vivo* activities

A great variety of adherent cells is rounded after intoxication. Only after chemical mutagenesis could a cell resistent to TcdB be obtained (Florin 1991). The cell has a defect in the UDP-glucose metabolism which results in a decreased UDP-concentration in the cytosol (Chaves-Olarte et al. 1996). In cell cultures the active concentrations (Eichel-Streiber et al. 1987; Popoff 1987; Martinez and Wilkins 1988; Ball et al. 1993) of the five LCT toxins are between 0.001 and 10 ng/ml (see Table 1). When used in vitro as an enzyme on lysed cells much higher concentrations have to be applied (Just et al. 1995). This points to the fact that the toxins are further processed and toxified, once they have reached the cytosol.

Applied intraperitoneally into mice, a single dose of the toxin acts lethal (see Table 1). This lethal effect was a major obstacle to getting TcdB-antibody responses in test animals (Eichel-Streiber, unpublished data). When given perorally in hamsters TcdA induces the disease and the hamsters finally die (Lyerly et al. 1985). In humans TcdB seems to be more important (Riegler et al. 1995). Probably both toxins act synergistically and are both needed for the onset of the disease.

Use in cell biology

The target GTPases of TcdA and TcdB are of major importance for the cell. CDC42, Rac, and Rho have been described as being involved in a cascade of reactions (Chant and Stowers 1995; Nobes and Hall 1995) connected with lamellipodia (CDC42), membrane ruffling (Rac), and stress fibre formation (Rho). Rac is involved in

oxidative burst of granulocytes (Ridley 1995). Rac and Rho play an important role in mast cell secretion (Price et al. 1995). Activated Rho induces PIP 5-kinase thus influencing the actin cytoskeleton and signal-transduction (Chong et al. 1994). Inactivation of Rho-GTPases by *C. difficile* toxins A and B leads to a decrease in phospholipase C and D activity due to a reduction in substrate supply (Schmidt et al. 1996a; Schmidt et al. 1996b). This is the first example for the use of *C. difficile* toxins in cell biology. These and the other LCTs will obviously soon become tools to dissect the interplay (signal transduction) between GTPases and the cell machinery.

References

Ball, D. W., van Tassell, R. L., Roberts, M. D., Hahn, P. E., Lyerly, D. M., and Wilkins, T. D. (1993). Purification and characterization of alpha-toxin produced by *Clostridium novyi* type A. *Infect. Immun.*, **61**, 2912–18.

Braun, V., Hundsberger, T., Leukel, P., Sauerborn M., and Eichel-Streiber, C. v (1996). Definition of the single integration site of the pathogenicity locus in *Clostridium difficile*. *Gene*, **181**, 29–38.

Chant, J. and Stowers, L. (1995). GTPase cascades choreographing cellular behaviour: movement, morphogenesis, and more. *Cell*, **81**, 1–4.

Chaves-Olarte E., Florin, I., Boquet, P., Popoff, M., Eichel-Streiber, C. v., and Thelestam, M. (1996). Low level of UDP-glucose in a mutant cell resistant to glucosyl-transferase toxins from *Clostridium difficile* and *C. sordellii. J. Biol. Chem.*, **271**, 6925–32.

Choe, S., Bennett, M. J., Fujii, G., Curmi, P. M. G., Kantardjieff, K. A., Collier, R. J., et al. (1992). The crystal structure of diphtheria toxin. *Nature*, **357**, 216–22.

Chong, L. D., Traynor-Kaplan, A., Bokoch, G. M., and Schwartz, M. A. (1994). The small GTP-binding protein Rho regulates a phosphatidylinositol 4-phosphate 5-kinase in mammalian cells. *Cell*, **79**, 507–13.

Depitre, C., Delmee, M., Avesani, V., L'Haridon, R., Roels, A., Popoff, M., et al. (1993). Serogroup F strains of *Clostridium difficile* produce toxin B but not toxin A. *J. Med. Microbiol.*, **38**, 434–41.

Dove, C. H., Wang, S. -Z., Price, S. B., Phelbs, C. J., Lyerly, D. M., Wilkins, T. D., et al. (1990). Molecular characterization of the *Clostridium difficile* toxin A gene. *Infect. Immun.*, **58**, 480–8.

Eichel-Streiber, C. v. (1993). Molecular biology of *Clostridium difficile* toxins. In *Genetics and molecular biology of anaerobic bacteria* (ed. M. Sebald), pp. 265–89, Springer Verlag, New York, Berlin.

Eichel-Streiber, C. v. and Sauerborn, M. (1990). *Clostridium difficile* toxin A carries a C-terminal repetitive structure homologous to the carbohydrate binding region of streptococcal glycosyltransferases. *Gene*, **96**, 107–13.

Eichel-Streiber, C. v., Harperath, U., Bosse, D., and Hadding, U. (1987). Purification of two high molecular weight toxins of *Clostridium difficile* which are antigenically related. *Microbial Pathogen.*, **2**, 307–18.

Eichel-Streiber, C. v., Suckau, D., Wachter, M., and Hadding, U. (1989). Cloning and characterization of overlapping DNA fragments of the toxin A gene of *Clostridium difficile. J. Gen. Microbiol.*, **135**, 55–64.

Eichel-Streiber, C. v., Laufenberg-Feldmann, R., Sartingen, S., Schulze, J., and Sauerborn, M. (1990). Cloning of *Clostridium difficile* toxin B gene and demonstration of high N-terminal homology between toxin A and B. *Med. Microbiol. Immunol.*, **179**, 271–9.

Eichel-Streiber, C. v., Warfolomeow, I., Knautz, D., Sauerborn, M., and Hadding, U. (1991). Morphological changes in adherent cells induced by *Clostridium difficile* toxins. *Biochem. Soc. Trans.*, **19**, 1154–60.

Eichel-Streiber, C. v., Sauerborn, M., and Kuramitsu, H. K. (1992). Evidence for a modular structure of the homologous repetitive C-terminal carbohydrate-binding sites of *Clostridium difficile* toxins and *Streptococcus mutans* glucosyltransferases. *J. Bacteriol.*, **174**, 6707–10.

Eichel Streiber, C. v., Laufenberg-Feldmann, R., Sartingen, S., Schulze, J., and Sauerborn, M. (1992). Comparative sequence analysis of the *Clostridium difficile* toxins A and B. *Mol. Gen. Genet.*, **233**, 260–8.

Eichel-Streiber, C. v., Meyer zu Heringdorf, D., Habermann, E., and Sartingen, S. (1995). Closing in on the toxic domain through analysis of a variant *Clostridium difficile* cytotoxin. *Mol. Microbiol.*, **17**, 313–21.

Eichel-Streiber, C. v., Boquet, P., Sauerborn M., and Thelestam, M. (1996). Large clostridial cytotoxins – a family of glycosyl-transferases modifying small GTP-binding proteins. *Trends in Microbiology*, **14**, 375–82.

Florin, I. (1991). Isolation of a fibroblast mutant resistant to *Clostridium difficile* toxins A and B. *Microb. Pathogen.*, **11**, 337–46.

Florin I., and Thelestam, M. (1986). Lysosomal involvement in cellular intoxication with *Clostridium difficile* toxin B. *Microb. Pathogen.*, **1**, 373–385.

Giry M., Arpin, M., Eichel-Streiber, C. v., Popoff, M. and Boquet, P. (1995). Transient expression of Rho A, B, C GTPases into Hela cells potentiates their resistance to *Clostridium difficile* A and B but not to *Clostridium sordellii* lethal toxin. *Infect. Immun*, **63**, 4063–71.

Green G. A., Schue V., and Monteil, H. (1995). Cloning and characterization of the cytotoxin L-encoding gene of *Clostridium sordelli*: homology with *Clostridium difficile* cytotoxin B. *Gene*, **161**, 57–61.

Henriques B., Florin, I., and Thelestam, M. (1986). Cellular internalization of *Clostridium difficile* toxin A. *Microb Pathogen.*, **1**, 455–463.

Hofmann, F., Hermann, A., Habermann, E., and von Eichel-Streiber, C. (1995). Sequencing and analysis of *Clostridium novyi's* α-toxin gene supports definition of a subgroup of clostridial toxins. *Mol. Gen. Genet.*, **247**, 904–13.

Hofmann, F., Rex, G., Aktories, K., and Just, J. (1996). The Ras related protein Ral is monoglucosylated by *Clostridium sordellii* lethal toxin. *Biochem. Biophys. Res. Commun.*, **227**, 77–81.

Johnson, J. L., Phelps, C., Barroso, L., Roberts, M. D., Lyerly, D. M., and Wilkins, T. D. (1990). Cloning and expression of the toxin B gene of *Clostridium difficile*. *Curr. Microbiol.*, **20**, 397–401.

Just, I., Fritz, G., Aktories, K., Giry, M., Popoff, M. R., Boquet, P., et al. (1994). *Clostridium difficile* toxin B acts on the GTP-binding Rho. *J. Biol. Chem.*, **269**, 10706–12.

Just, I., Selzer, J., Wilm, M., Eichel-Streiber, C. v., Mann, M., and Aktories, K. (1995). Glycosylation of Rho proteins by *Clostridium difficile* toxin B. *Nature*, **375**, 500–3.

Knoop, F. C., Owens, M., and Crocker, I. C. (1993). *Clostridum difficile*: clinical disease and diagnosis. *Clin. Microbiol. Rev.*, **6**, 251–65.

Krivan, H. C. and Wilkins, T. D. (1987). Purification of *Clostridium difficile* toxin A by affinity chromatography on immobilized thyroglobulin. *Infect. Immun.*, **55**, 1873–7.

Krivan, H. C., Clark, G. F., Smith, D. F., and Wilkins, T. D. (1986). Cell surface binding site for Clostridium enterotoxin: evidence for a glycoconjugate containing the sequence Gal alpha 1-3Galbeta 1-4GlcNAc. *Infect. Immun.*, **53**, 573–81.

Lyerly, D. M., Saum, K. E., MacDonald, D. K., and Wilkins, T. D. (1985). Effects of *Clostridium difficile* toxins given intragastrically to animals. *Infect. Immun.*, **47**, 349–52.

Lyerly, D. M., Phelps, C. J., Toth, J., and Wilkins, T. D. (1986). Characterization of toxins A and B of *Clostridium difficile* with monoclonal antibodies. *Infect. Immun.*, **54**, 70–6.

Lyerly, D. M., Krivan, H. C., and Wilkins, T. D. (1988). *Clostridium difficile*: its disease and toxins. *Clin. Microbiol. Rev.*, **1**, 1–18.

Lyerly, D. M., Barroso, L. A., and Wilkins, T. D. (1991). Identification of the latex test-reactive protein of *Clostridium difficile* as glutamate dehydrogenase [see comments]. *J. Clin. Microbiol.*, **29**, 2639–42.

Martinez, R. D. and Wilkins, T. D. (1988). Purification and characterization of *Clostridium sordellii* hemorrhagic toxin and cross-reactivity with *Clostridium difficile* toxin A (enterotoxin). *Infect. Immun.*, **56**, 1215–21.

Meador III, J. and Tweten, R. K. (1988). Purification and characterization of toxin B from *Clostridium difficile*. *Infect. Immun.*, **56**, 1708–14.

Müller, F., Stiegler, C., and Hadding, U. (1992a). Monoclonal antibodies specific for *Clostridium difficile* toxin B and their use in immunoassays. *J. Clin. Microbiol.*, **30**, 1544–50.

Müller, H., Eichel Streiber, C. v., and Habermann, E. (1992b). Morphological changes of cultured endothelial cells after microinjection of toxins that act on the cytoskeletons. *Infect. Immun.*, **60**, 3007–10.

Nobes, C. D. and Hall, A. (1995). Rho, Rac, and Cdc42 GTPases regulate the assembly of multimolecular focal complexes associated with actin stress fibers, lamillipodia, and filopodia. *Cell*, **81**, 53–62.

Popoff, M. R. (1987). Purification and characterization of *Clostridium sordellii* lethal toxin and cross-reactivity with *Clostridium difficile* cytotoxin. *Infect. Immun.*, **55**, 35–43.

Popoff M.R., Chaves-Olarte, E., Lemichez, E., Eichel-Streiber, C. v., Thelestam. M., Chardin, P., Cussac, D., Antonny, B., Chavier, P., Flatau, G., Giry, M. and Boquet, P. (1996). Ras, Rap and Rac are the target GTP-binding proteins of *C. sordellii* lethal toxin glucosylation. *J. Biol. Chem.*, **271**, 10217–24.

Pothoulakis, C., Barone, L. M., Ely, R., Faris, B., Clark, M. E., Franzblau, C., et al. (1986). Purification and properties of *Clostridium difficile* cytotoxin B. *J. Biol. Chem.*, **261**, 1316–21.

Price, L. S., Norman, J. C., Ridley, A. J., and Koffer, A. (1995). The small GTPases Rac and Rho as regulators of secretion in mast cells. *Curr. Biol.*, **5**, 68–73.

Ridley, A. J. (1996). Rho: theme and variations. *Curr. Biol.*, **6**, 1256–64.

Riegler, M., Sedivy, R., Pothoulakis, C., Hamilton, G., Zacherl, J., Bischof, G., et al. (1995). *Clostridium difficile* toxin B is more potent than toxin A in damaging human clonic epithelium *in vitro*. *J. Clin. Invest.*, **95**, 2004–11.

Rolfe, R. D. and Song, W. (1993). Purification of a functional receptor for *Clostridium difficile* toxin A from intestinal brush border membranes of infant hamsters. *Clin. Infect. Dis.*, **16**, (Suppl. 4), S219–27.

Rupnik, M., Braun, V., Soehn, F., Hofstetter, M., Janc, M., and Eichel-Streiber, C. v. (1996). Lack of detection of *Clostridium difficile* toxins due to genetic polymorphism of the toxic genes. *Zentralbl. Bakteriol. Suppl.* (accepted for publication).

Sauerborn, M. and Eichel-Streiber, C. v. (1996). Interaction of toxin A of *C. difficile* with its cell receptor. *Zentralblatt. Bakteriol. Suppl.* (accepted for publication).

Sauerborn, M., Hofstetter, M., Laufenberg-Feldmann, R., and Eichel-Streiber, C. v. (1991). The C-terminal repetitive domain of *Clostridium difficile* toxin A and its significance